IRONIC TECHNICS

Other books from Automatic Press ♦ $\frac{V}{I}$P

Formal Philosophy
edited by Vincent F. Hendricks & John Symons
November 2005

Masses of Formal Philosophy
edited by Vincent F. Hendricks & John Symons
October 2006

Political Questions: 5 Questions for Political Philosophers
edited by Morten Ebbe Juul Nielsen
December 2006

Philosophy of Technology: 5 Questions
edited by Jan-Kyrre Berg Olsen & Evan Selinger
February 2007

Game Theory: 5 Questions
edited by Vincent F. Hendricks & Pelle Guldborg Hansen
April 2007

Legal Philosophy: 5 Questions
edited by Morten Ebbe Juul Nielsen
October 2007

Philosophy of Mathematics: 5 Questions
edited by Vincent F. Hendricks & Hannes Leitgeb
January 2008

Philosophy of Computing and Information: 5 Questions
edited by Luciano Floridi
Sepetmber 2008

Philosophy of the Social Sciences: 5 Questions
edited by Diego Ríos & Christoph Schmidt-Petri
September 2008

Epistemology: 5 Questions
edited by Vincent F. Hendricks & Duncan Pritchard
September 2008

Probability and Statistics: 5 Questions
edited by Alan Hájek & Vincent F. Hendricks
November 2008

See all published and forthcoming books at
www.vince-inc.com/automatic.html

IRONIC TECHNICS

Don Ihde

Automatic Press ♦ $\frac{V}{I}$P

Automatic Press ♦ $\frac{V}{I}$P

Information on this title: www.vince-inc.com/automatic.html

© Automatic Press / VIP 2008

This publication is in copyright. Subject to statuary exception and to the provisions of relevant collective licensing agreements, no reproduction of any part may take place without the written permission of the publisher.

First published 2008

Printed in the United States of America
and the United Kingdom

ISBN-10   87-92130-18-6   paperback
ISBN-13   978-87-92130-18-1   paperback

The publisher has no responsibilities for
the persistence or accuracy of URLs for external or
third party Internet Web sites referred to in this publication
and does not guarantee that any content on such
Web sites is, or will remain, accurate or appropriate.

Typeset in $\LaTeX 2_\varepsilon$
Photo and graphic design by Vincent F. Hendricks

Contents

Introduction	iii
1 Stupidity in the Knowledge Society	1
2 The Designer Fallacy and Technological Imagination	19
3 Aging: I don't want to be a Cyborg	31
4 Of Which Human are we Post?	43

Introduction

The issues I am addressing in *Ironic Technics* are all very contemporary: the 'knowledge society', 'designer control of technologies,' humans as 'cyborgs,' and questions of 'posthumanity.' It is often noted that irony is a favored rhetorical style in postmodernity. Presumably one reason for this is because our era is one in which the old certitudes have been called into doubt. Irony thus seems to fit into a culture of doubt and the re-questioning of our older beliefs and values. In my case my taste for irony has had a long and personal history. Existentialism was an intellectual fashion during my undergraduate days with most of us reading Albert Camus, Franz Kafka, Jean Paul Sartre, Friedrich Nietzsche among others. My favorite became Søren Kierkegaard, clearly a master of irony. His modes of irony took different forms, sometimes comical as in the figure who trying to convince himself that he is not insane, places a ball in his back pocket so that when he walks and the ball hits his buttock when striding along, he says to himself, "I am not insane. I am not insane...." And sometimes the irony carries itself in a most serious way, as in Kierkegaard's 'phenomenological variations' on Abraham's doubts about sacrificing Isaac in *Fear and Trembling*. Abraham, chosen of God, after seventy years has Isaac by his aged wife, Sarah, only to be tempted by God with a command to sacrifice Isaac on Mt. Moriah. Kierkegaard poses a range of possibilities: could Abraham think God is mocking him? Should he, on arrival at Mt. Moriah, substitute himself as the sacrifice? Or, what if he had spotted the ram before God spoke and thus made the substitute himself without faith? (And, I would add to the text the later variation of Woody Allen, why should Abraham take the word of someone who speaks from the heavens with a loud voice?) I avidly read all the Kierkegaard I could find, sometimes many times over.

Later, while attending theological school, I tried to emulate Kierkegaardian irony in the form of a pseudonymous newsletter, "The Gadfly," in which I mocked overly serious and pious practices and sometimes hypocritical claims by the hyper-religious, rather like Kierkegaard's take upon the Hegelians of his time. And still later, this time while serving as a Dean of Humanities and

Arts in my university, I repeated the exercise with "The Humanities Gadfly," another pseudonymous newsletter. This was a time when the 'business model' of academic administration was becoming popular—our then Provost wanted each unit, including philosophy, to become a 'profit center,' generating monies for its own operations. It was clear that the humanities and arts were severely disadvantaged in such a situation since research grants which were deemed one sort of 'profit' were hard to come by for humanists. Never plush, but in the time of Thatcher in the UK and Reagan in the US, resources for the arts and humanities were limited and restricted. So, in one of my Gadfly issues, I argued that a fair comparison of humanities and the sciences had to include "handicapping odds." As it turned out at that time, the average 'win' of a National Science Foundation grant had odds of 1/3, but a National Endowment for the Humanities Fellowship had odds of more than 1/30, and since a good number of us in philosophy had won such fellowships, I was able to claim that with handicapping odds taken into account, our philosophers were doing better than our physicists! At the least the "Gadfly" generated lively discussion within the College, although most readers had discovered who the actual author was.

What does a preferred rhetorical style have to do with technics? The connection is not simply a matter of personal taste. I had, even in my earliest post-Ph.D days, already begun to think about technologies, and by the 70's to write about human-technology relations. My 1979 book, *Technics and Praxis: A Philosophy of Technology,* explored the various roles and relationships humans employed in technological contexts. But unlike Martin Heidegger, to whom I had dedicated the book, I instead undertook the examination, not of the *essence* of Technology, but of specific technologies. What these taught me was how diverse they were, how differently embedded in different cultures even the same technologies may be, and above all, how technological history is so full of surprises, the unexpected and with unintended and unpredictable side-effects. There was bound to be in such studies, grist for much irony.

Here the topic is *technics,* that is the realm of human-technology inter-relations which are connected to hopes, fears, imaginations, today focused upon the new and more complicated, higher technologies which relate the topics noted above. The chapters originally were invited lectures or essays in a Euro-American context. The first chapter, "Stupidity in the Knowledge Society" was first given at a conference on the knowledge society at the university in Aarhus, Denmark, 2005. Denmark was considering its own "branding" which was to include the designation, a knowledge society. Peter Drucker had claimed, in 1994, that the biggest social transformation of the century was to have moved from an industrial to a 'knowledge' society. Of course such a claim carried implicit utopian implications which not much later had the dark cloud of contemporary terrorism cast upon it. Terrorists, too, use the internet and the tools of a knowledge society. One of the traditions of modernism is the persistent claim that some "revolution" has or is occurring. Does the knowledge society replace the industrial society?

My second chapter, "The Designer Fallacy and Technological Imagination," was originally an essay for a collection on technological literacy, initiated by John Dakers, later expanded as a presentation for the Society for Philosophy and Technology (the version I reprint here). There is today much discussion of the design process, often in terms which presume that designers can build into their designed technologies, the uses for which they were intended. Again, drawing on a rich set of historical examples, it often turns out that the ultimate uses of technologies are either very different, or vary multiply from such designer intent—another irony for technics. Does the old model of inventive humans imaginatively designing technologies, which then fulfill the designer's intended purposes hold?

The third essay was invited as a contribution to a special issue on cyborgs in *Phenomenology and the Cognitive Sciences* "Aging: I don't want to be a Cyborg." Drawing upon both historical examples and my own experience of aging, I look at a series of human body-transforming technologies with the various ironies which crop up in this genre of human-technology-animal hybrids. In the version printed here, I expand the original essay to cover my own case following heart surgery, the contemporary style for which high technologies are routinely employed.

Fourthly, "Of Which Human are we Post?" was first presented at a conference on posthumanism at the University of Nottingham, UK, 2007, and later for the Templeton program at Arizona State University, 2008. Here, again, there are echoes of utopianism from both post and transhumanism currents. Behind much posthuman thinking is the notion that technics can improve upon our current being, a notion which I seek to re-frame by returning it to the inter-relation between humans and technologies.

While the tone of irony runs through all the chapters, which seek to avoid both utopian and dystopian interpretations of technics, there is a subtheme which owes a debt to a fellow thinker who clearly had his own take on the ironic: Otto L.Bettmann, founder of the Bettmann archive in New York. Bettmann had been the curator of rare books at the Prussian Art Library in Berlin before emigrating to America in 1935 when so many other Jewish-German intellectuals also fled National Socialism. He became a collector of graphic, photographic and other visual arts—the Bettmann archive has over eleven million such prints and photos. It is today owned by Bill Gates. In his *The Good Old Days— They Were Terrible* (Random House, 1974) he compiles graphics concerning life in 19th century America, dealing with air, traffic, housing, health and the like. The vivid cartoons, photos, reveal his particular and deep sense of irony. Rather than the "good old days" as simpler, kinder and better than today, his visual journey debunks such nostalgia. New York City traffic jams, as well as those in Chicago and elsewhere, were as horrendous as the worst of today's—except the streets were filled with horse drawn vehicles, trolleys, and other pre-gasoline vehicles. And the horse-induced accident death rate was *ten times higher per capita* than today's automobile accident rates. Air pollution, water pollution, lack of sanitation is illustrated vividly, including the parade of garbage barges traveling just outside the swimming areas of New York and New Jersey beachs. He has often inspired me to avoid all romanticism and dystopianism which so filled the early philosophies of technology in the early twentieth century. Irony, I claim, has a critical effect, urging multiple perspectives for all examinations of technics.

Irony may be the style which unites these essays, but there is also a *reconstructive* aim which accompanies the critique: I argue that technologies are *multistable,* that while the materiality of the different artifacts display constraints in relation to human users, monolithic uses are impossible. Thus prediction is complicated and the proliferation of variable trajectories can always occur. I also argue for an 'inter-relational' ontology, whereby the human-technics situations are polymorphic and plastic. If we humans 'invent' technologies; then reciprocally, our technologies re-shape our lifeworlds and thus us within these worlds as well.

Acknowledgements: My long term travels have often been initiated via internet acquaintance before actual face-to-face meetings and that is the case for the essays here as well. Now friends as well as professional colleagues, I acknowledge Finn Olesen from Aarhus University, Denmark, as the inviter for chapter one; John Dakers of the University of Glasgow, UK, and Pieter Vermaas of Delft University, the Netherlands, as the inviters for chapter two. Evan Selinger, my former student and co-editor of our *Chasing Technoscience* (Indiana, 2003), Rochester Institute of Technology, USA, called for the cyborg chapter, and Chris Johnson, Nottingham, UK, and Hava Samuelson, Arizona State University, USA, invited the essay on posthumanism.

Chapters one and four have not be previously printed, and while chapter four appeared online with the Templeton The Global Spiral, both are print original to this volume. Chapter two, in the version re-printed here, originally appeared in *Philosophy and Design: From Engineering to Architecture,* edited by P.S. Vermaas, P. Kroes, A. Light, S.A. Moore (Springer, 2008) reprinted here by permission of Springer Publishers. Chapter three, in a shorter version, originally appeared in *Phenomenology and Cognitive Science,*Vol 7, No.3 special issue on Cyborg Embodiment: Affect, Agency, Intentionality and Responsibility, edited by Evan Selinger and reprinted here by permission of Springer Publishers.

To list all those who contributed to discussions which helped shape these essays would occupy too many pages for an introduction. But I want to acknowledge my close colleague and often collaborator, Robert Crease, Stony Brook University. He and I have had many exchanges on these matters and he often directs the technology research seminar while I am traveling. The seminar in which trial presentations and critical discussions occur, is the locus of much interchange of matters of technics. The participants, Visiting Scholars from many countries, doctoral students from several departments, and faculty colleagues are all part of this group of interlocutors.

My son, Mark, and I discuss philosophy at home and commuting back and forth from home to campus, and my wife, Linda, provides support by insightful editorial suggestions, by encouraging and tolerating the travels which make international discussion possible as well, in spite of the burdens which such absences cause.

1
Stupidity in the Knowledge Society

What is the *knowledge society?* That was my first question upon preparing for this occasion. So, to find out, I did what most contemporary people might do as an initial move: I turned to the internet, typed in "the information society," and the first entry that popped up was Peter Drucker's 1994 Edwin L. Godkin Lecture to the John F. Kennedy School of Government at Harvard University. His title: "Knowledge Work and Knowledge Society: The Social Transformations of this Century."

It seems then that 'the knowledge society,' is one of many 'societies'—and 'posts' which have sprung up at the end of the 20th and beginning of the 21st centuries. Alvin Toffler coined "the information society" in his 1980 *The Third Wave;* we've been in a post-industrial society for a while, too, according to some; and there are other posts as well—post Cold War, postmodernism, and even my own 'postphenomenology.' What all these terms imply, is that some change has occurred, which is new and dramatic, and which outdates what preceded the new society or post-condition. Now, frankly, this notion of 'revolution' and the 'new' is actually a continuation of *modernity* itself. Modernity made "revolutions" normative. From Descartes' method of doubt in the 17th century, to Kant's 'Copernican Revolution' in the 18th century to Kuhn's 'Structure of Scientific Revolutions' in the 20th, revolutions are simply the sedimented *tradition* of modernism itself. Thus we must be cautious here.

I begin with excerpts from Drucker: "No century in human history has seen such radical and swift social transformations as the twentieth century...."[1] So, here we have another modernist 'revolution'—but with an empirical claim: "In the first decades of this century up to the first world war society ....even in the most highly industrialized ones...was in its structure still pretty much what it had been since the first humans became farmers and

---

[1] Transcript: The 1994 Godkin Lecture by Peter Drucker, May 4, 1994. http://www.ksg.harvard.edu, p.1

settlers on the land some five thousand years earlier."[2] Drucker observes that a rural population, specifically farmers, "Even in the UK and Belgium...were still the largest single group in the population...in the developed countries, including the US, they were still close to being an absolute majority."[3] He was right, and I shall return to this. But after world war II, Drucker sees the emergence of a *knowledge society* with a rapidly growing population of *knowledge workers*. This population, "gains access to work, job and social position through formal education."[4] The "knowledges" which Drucker pluralizes, are necessarily specializations, of different types, but all requiring formal educational training, thus making schools of many types necessary in the formation of knowledge work. Drucker sees open access to such education as an entry to knowledge work; but also knowledge workers are *employed*, and thus 'workers.' I have let Drucker set the stage. But I also want to point out that Drucker's version of the knowledge society emphasizes social changes and roles with technologies at most in the background. In contrast, I will foreground the relationships with technologies. This so that whatever shape knowledge takes in the knowledge society, its mediation will include technologies of mediation.

In my second internet attempt, I added "Denmark" to the knowledge society, and along with Norway, the UK and several other northern European countries, the notion that they are 'knowledge societies' belongs to their self 'branding' process. (I haven't noticed that tendency in my own country.) But, I must now be very careful with my title, "Stupidity in the Knowledge Society"!

I admit, the title is at first strange. But, unless you are a rabid utopian, you cannot believe that any 'revolution' or new stage will succeed in eliminating stupidity. And while I shall not argue that a knowledge society succeeds in reducing or increasing stupidity, I shall argue that the shapes and forms of stupidity, and for that matter all knowledge, *is relative to the modes of communication and technologies employed in whatever type of society it may be.*

I shall begin with a series of vignettes, or disparate pieces of 'knowledges' about ideas, events, and thoughts which will, at first, seem quite unconnected to the knowledge society:

---

[2] *ibid.*, 2.
[3] *ibid.*, 1.
[4] *ibid.*, 2.

1. Stupidity in the Knowledge Society   3

- To the best of my knowledge, no Danish cartoonist bombed the Golden mosque in Samarra, Iraq.

- A few years ago the following people were on a panel at New York University, discussing the impact of technologies upon society: Alain Finkelkraut, a French 'Heideggerian' who had just brought out a book; Regis Debray, a socialist thinker from the days of 1968, Paris; Neil Postman, dystopian author concerning media culture, Jaron Lanier, red-headed, dreadlocks hypermodernist who coined the term, "virtual reality," but who is also a reader of Heidegger;– and me. We were seated 'right' to 'left' with the first three mentioned on the right; Lanier and me on the left. The debates were hot and the four hundred people in attendance kept the question session going until midnight—but the highpoint for me came from Lanier, who observed that Heidegger's notion of 'modern technology' was really that of 'rustbelt, industrial technology' which enframes the world [*Gestell*] and challenges nature as 'standing reserve' [*Bestand*], and not that of contemporary electronic-information technology which –at the least—is unlike the older technology, with the implication that the latter would replace the former.

- Remember the Cold War? The two large groupings of Russia and the East versus the United States and NATO threatening each other with MAD [Mutual Assured Destruction] via nuclear arsenals. The notion was that large,'responsible' nations with such a balance of terror would not possibly launch an attack. It worked in that during the several decades of the Cold War, no nuclear war occurred—but was it because of the theory, or due simply to lucking out? There were close calls: the Cuba Missile crisis, several recently revealed wargame excercises momentarily mistaken as real imitating the movie "War Games,", and other close calls as well. Part of the idea was that such a stand-off equalized the parties, a democracy of terror as it were.

- During that time I proposed a *Gedankenexperiment* which takes the MAD idea to a *reductio ad adsurdum:* why not bury a bomb big enough to explode the entire planet and then give every single individual human a remote control device such that anyone could detonate the device. Would this not be the ultimate democracy, making every single individual equal in power and thus prompt total rationality and

1. Stupidity in the Knowledge Society

security by the strange logic of MAD, now individualized? Not a single person hearing of this idea ever thought this outcome likely, since there was bound to be some suicidal maniac who would go ahead and detonate the world.

- But, if individuals are not to be trusted, what about groups—or nations—aren't these more stable? Yet, historically, from Masada to the Jones cult in Guiana, groups too, have sometimes resorted to group suicide. And since the Cold War, North Korea, Pakistan and India have gone nuclear, with Iran possibly next—but we should also not forget that after the Cold War, 12 countries abandoned their nuclear arsenals, including such unexpected ones as South Africa and Libya.

- And what about 'WMDs' or weapons of mass destruction? There weren't any in Iraq, yet war was launched, but there were then and are now many 'WSDs', weapons of simple destruction—home made bomb belts with ball bearings or nails and willing suicides—these, with IEDs [Improvised explosive devices, again simple and easy to make] continue to keep the world unsafe. These have multiplied across the world and through our media we are all too aware of their results. There have been over three-thousand such WSD events since the first of this year.

You may now ask, what has all this to do with the knowledge society? Be patient and I will eventually weave it into an answer:

I begin with Lanier's suggestion which actually contains two theses: is the technology of the 'knowledge society' qualitatively different from that of 'rust belt industrial' technology? And, in an implied modernist 'revolution' sense, does this imply that we are moving away from the latter toward the former? And, if this is so, then, has Heidegger been superceded? The answer to the first question is clearly a qualified 'yes.' Here are a few indicators:

- Industrial technologies often have a trajectory towards gigantism, the mega-technologies involved with processes like mining, factories, monumental engineering processes, and if we add military technologies, we add massive destructive forces. These technologies are often associated with massive consumption of energy and resources. In contrast, electronic technologies associated with the production or dissemination

1. Stupidity in the Knowledge Society 5

of information often have trajectories towards miniaturization and low energy and resource consumption. For example, to produce aluminum, which entails mining, reduction to a metal, and processing, one requires 507 *billion BTU's* to produce 86 billion tons of aluminum, with one-third the cost of the end product the cost of energy. In contrast, when hooking up my previous mountain home to solar power, we asked ourselves what we really wanted—lighting, computer-printers, mobile phones and electronic CD players, etc. – we found we did not even need an AC inverter and settled for a simple six panel + battery set up which delivered 12 volt DC and supplied all these needs! All running at the same time amounted to less energy being used than that of a hair dryer.

- Similarly, earlier industrial technologies 'consumed' massive amounts of human labor, often re-patterned into Taylorized, de-skilled processes such as assembly lines, which provoked worker dissatisfaction and fed the thoughts of those arguing that industrial capitalism 'alienated' labor, both from one's work and from nature. Today's electronic entertainment, communication, and information technologies seem to take a different direction by placing users into connections with each other, providing play time, and even empowering people in different ways. For example, I would not be able to be here doing this lecture were it not for email and the internet.

- Furthermore, the industrial technologies called for concentrations of labor, bringing workers from afar to a central location and into tasks largely of a single purpose—to make automobiles along an assembly line; whereas miniaturized electronic technologies seem distributive, carried anywhere, and are frequently *multi-tasked*. The mobile phone which takes pictures, reads bar-codes, and connects to email an example.

Were these generalizations to be taken as true, it might then seem that the shift from an industrial to knowledge style technologies would clearly be a good direction to go. But, there are some serious catch 22's to this portrayal and these include serious doubt about any demise or replacement of industrial technologies:

- First, it should be obvious that electronic technologies such

1. Stupidity in the Knowledge Society

as are used in knowledge practices have not *replaced* industrial technologies, but rather have been *added on* to these as a different layer of technologies as it were. The same guys who operate the mega-technologies carry mobile phones in their five story high earth movers. These earth movers are designed by Krupps, the same company which brings you coffee makers.

- Admittedly, some information technologies can be placed into use in what *seems* to be a leapfrog phenomenon, i.e., use without going industrial, but this is largely illusory since, while the Masai herding cattle in Africa also use cell phones, these belong to a vast system of microwave towers and energy supplies needed to have this connective technology work. Nonetheless, the user is potentially linked, globally, to virtually anywhere.

- And, the mega-technologies have also changed—many no longer require centralized labor—the offshore oil platform is staffed by relatively few people and is located far from any factory or city. Even many factories, including auto assembly plants, operate with more robots than human hands. This is no longer like Henry Ford's early labor intensive process.

- And, I wouldn't be here today except for an atmosphere polluting, fossil fuel burning jet plane either.

So, the answer to the second question seems to be "no," a knowledge society is not 'replacing' an industrial one, but it does play a role within, or in addition to an already industrialized world. In this sense the modernist notion of revolution as replacement fails. And, it fails in part because the *predictions* implied by replacement theories fail. In a recent book, *Technology Matters,* David Nye points out that George Wise did an interesting study of technology predictions by examining 1500 such predictions made at the turn of the 20th century by engineers, historians, and others. Of these, only one third even came vaguely to be; one third were falsified and one third have remained in ambiguity. Thomas Hughes, making a similar observation, noted that *futurologists* have virtually the *worst* predictive powers. And, finally, Edward Tenner specifically lists some of the worst unexpected results of predictions—such as the now infamous one that the electronic revolution would produce a *paperless society*.⁵ But my favorite relating to the information society is the comment by Tom Watson of IBM in 1958, "I think there is a world market for about five computers."⁶

I guess that part of what I am trying to do, is to wean us from the current hype and tendency to overgeneralize and romanticize the new. Might we, for example, have a too one-dimensional and a too short-range view of industrialization itself and its related effects? In September, 2005, *Scientific American,* published a special issue on "Crossroads for Planet Earth," which argued that by the 21st century the human race had reached a number of crucial crossroads. Now, I am often skeptical of the *Scientific American* because it tends to be too optimistic about what science can do for us, but in this case the statistical analyses about long term trends were often compelling, particularly about the results of the long range impact of industrialization:

- At first, beginning with the 18th century, there was an enormous increase of population growth, due in part to improved sanitation and medical practices as well as to increased food production capacities related to industrialization and its modern scientific knowledges. This growth began to accelerate two centuries ago, but reached its climax in the 20th century, as *Scientific American* put it, "no person who died before 1930 had lived through a doubling of the human population."⁷ And, within this explosion a significant sub-feature also emerged—one was the shift to an increasingly *urban* population, "Until approximately 2007, rural people have always outnumbered urban people. From 2007 forward, urban people will outnumber rural people."⁸ When I was born, there were 2.16 billion people on the earth, compared to 6.5 billion today. And from my own experience as a Kansas farm boy, I, myself have witnessed the shrinkage of actual farm inhabitants in the US to its present 3% of the total US population, with the rural/urban shift going from 55% urban at my birth to 85+% today and increasing. But

---

⁵Edward Tenner, *Why Things Bite Back: technology and the revenge of unintended consequences* (New York: Alfred A. Knopf 1996), p. ix.

⁶New York Telephone.

⁷*Scientific American,* special issue, "Crossroads for Planet Earth," September 2005, p. 48.

⁸*ibid.*, p. 48.

the peak of doubling occurred between 1960, with 3 billion inhabitants, to 2005 with 6.5 billion inhabitants. This human population expansion, worried about by Malthus in the 19th century, and worried even more about by 20th century "population explosion" theorists, did not, however, last.

- One can now cogently argue today that the leveling off of the population, world-wide, is even more dramatic than the earlier increase. Again, the *Scientific American,* "Nor is any person born in 2050 or later likely to live through a doubling of the human population."[9] The absolute peak was passed in 1965-1970, with a 2.1 % growth rate, and has already fallen to today's 1 %, and with the most industrialized countries—discounting immigration—now virtually all below self-replication levels. One of the most startling facts about populations is that once countries reach a certain level of industrialization—no matter what the religion, what the culture or the location—this leveling effect comes into play. Families in both the USA and Japan displayed this shift after 1945 and both countries today—discounting immigration—are below self-replication levels. This rapid shift is also accompanied by a new sub-feature. *Scientific American* observes, " Before 2000, young people always outnumbered old people. From 2000 forward, old people will outnumber young people."[10]

To summarize, it appears that the effects of industrialization, then, led to a rapid explosion of the human population with a shift from rural to urban balances; followed by a leveling off of a population with a shift from a younger to an older population. The first phase of this effect is older and better known than its second phase which is only now coming into full view.

A second major effect of industrialization, *Scientific American* claims has to do with the reduction of extreme poverty. By extreme poverty is meant starvation, totally inadequate medical care, and lack of income sufficient to provide basic needs. The current claim is that today 5.5 billion of earth's 6.5 billion people have escaped extreme poverty, and that percentage has been reached only since the later Industrial Revolution, but mostly since the 1980's. The most recent and dramatic shrinkage relates to the

---

[9] *ibid.,* p. 48.
[10] *ibid.* p. 48.

1. Stupidity in the Knowledge Society   9

rapid industrialization in Asia such that the desparately poor today are mostly concentrated in Africa and the most rural areas of Asia and Latin America.

Interestingly, a third massive effect was left out by the *Scientific American*—education and literacy was also a massive effect of industrialization with literacy rates well over 90% in all industrialized countries. It is this massive effect which maps most closely upon the notion of the knowledge society. And, this third effect is what Drucker focuses upon in his notion of the knowledge society.

The one negative factor resulting from industrialization and this is the factor which poses the 'crossroads' issue, is of course, the drastic use of natural resources with environmental and pollution effects leading to global warming. That is *the issue* which faces late industrialization. And, it may be the crucial factor which transcends both industrial and knowledge 'societies.' So, regarding the Lanier observation, I would answer "yes' to the question about a qualitative difference between industrial and knowledge technologies, but "no" to any significant factors showing that knowledge technologies will simply *replace* industrial ones. (But, this does not mean that industrial technologies cannot themselves be improved or change. More on these points later.)

So let us turn now more specifically to the technologies which support the knowledge society. Almost all electric powered such technologies began to occur only at the end of the 19th century, with the telegraph in place by the 1880's, followed in the 20th century with telephone, undersea cables, and radio technologies. Most were analog technologies, but they were also much faster and farther reaching than the slow paper based communications which preceded them. Most were also language based, although sometimes reduced to codes as per the telegraph. Similarly, the roots of audio-visual displays also took place in this time frame, including cinema which, with film, exceeded the spatial-temporal bounds of earlier magic lantern and home stereoscopic vision sets. High speed and transportability point to the globalizing effects of knowledge technologies. Knowledge society technologies are fast, transportable, connective and thus *global*.

But what today we might recognize as the major factor producing a knowledge revolution rightfully may be associated with computer process technologies. Early electronic computers began as large, bulky room-sized processors whose first use in massive data computation was associated with the Manhattan Project, the making of the atomic bomb near the end of World War II.

10   1. Stupidity in the Knowledge Society

By the 40's problems which would take an army of mathematicians decades, could be handled by computers, and even the first simulations, Monte Carlo simulations, were used in that terrible project. If we analyze this event, we can see that what was happening was that humans, now related to powerful, technological means of calculations, were able to create the awesome power of the atom bomb, later the hydrogen bomb. Still later, as these same *knowledge society* participants began to dream of 'peaceful uses,' they began to fantasize gigantic harbors to be created by burying atom bombs and exploding them to produce vast craters as a harbor base. Similarly, they fantasized that even the weather would be controlled by 'blowing up' hurricane funnels. These same hubris laden thinkers are planning to engineer earth to prevent global warming. These projects, actually discussed by the same intellectuals who produced the bomb, were conceived of in the fifties—and although retrospectively we may be appalled by this naivete combined with hubris, we also begin to glimpse *stupidity* playing a role in the very construction of the beginnings of the knowledge society. The intellectuals involved, premier denizens of the knowledge society, perfectly fulfilling the Drucker criteria, were figures such as Herman Kahn, Edward Teller, and even John von Neumann, most associated in one way or another with MIT. Here, then, is a history in which a powerful atomic technology, made possible by a powerful computational technology, synthesizes the 'industrial' with 'knowledge' in a fateful way. Yet, today, the single primary use of supercomputers, located in Japan, is to produce simulations and models of global weather with our well known need to understand global warming.

The second related and computer-connected technology necessary for a knowledge society, is, of course, the internet. And those of you who are technologically literate, know that this technology was also born, not from World War II, but in the midst of the Cold War. Under the threat of nuclear devastation, some means of 'secure,' and non-centralized, secret communications were needed. At first restricted to the military and security-cleared university personnel, the internet was built to survive any pre-emptive strike by a hostile power. And although this technology was more 'defensive' than offensive, it, too, was a 'wartime' technology. Yet, once unleashed from its design-intended bounds of a military-knowledge society set of users, it, too changed. Far from being secure, we are now all too familiar with 'hacking,' 'identity theft,' and 'spam' and the wars to protect ourselves from such intrusions.

Virus programs, scam mail, spyware protection, and spam blockage are also part of the knowledge society everyday world lifeworld and are displays of another kind of stupidity which permeates a knowledge society.

We are now ready to return to two of my other seemingly unconnected vignettes: MAD, itself *mad,* was dreamed up precisely by knowledge society Cold Warriors. Convinced, as many still are, that humans and societies operate by 'game theories,' which are often analogs to computer programs which today serve as our *epistemology engines,* that is the models by which we think we can simulate psychology, economics and in this case, wars, MAD simply was an extension of this style of knowledge society thinking. Somehow, we survived in spite of this. In this, Heidegger has not been superceded: this is the calculative society in supreme example.

Equally, fortunately, no one took up my ultimate democracy thought experiment—at least not with respect to MAD. But, ironically, we are now very close to such a 'democracy' in a new context, that of the world growth of the internet, to which I shall now turn. In my earlier thought experiment, everyone was connected with a remote to a planet-destructive device—that did not happen. But the dream today is that everyone can (and should) be connected to the 'net in a democracy of communication and knowledge.

This has not happened....yet. But the growth rate is phenomenal. Already one in six is connected. And, if the internet distribution is mapped it would largely once again replicate the previous maps with concentrations in the most developed countries, but one must also remember that in the less developed parts of the world, the internet café rather than the home user, is likely to be the dominant model, and thus there are more users than simply machines. How else could I receive so many Nigerian scams? And this trajectory does map upon my ultimate democritization thought experiment: what will things look like when *everyone* can go online? [Here, of course, I must take my own cautions to bear and not fall into a futurist trap of wild extrapolation.]

So, the task for the new thought experiment is: what will the internet as a knowledge society technology be like *if and when everyone is on?* In general, the answer must include whatever anyone can say, display, do, communicate, receive, etc. will be on. And we have a pretty good idea about this from the current practices upon the internet.

1. Stupidity in the Knowledge Society

- First, as I noted in the beginning, my own first move was to the 'net to find out what was there about the 'knowledge society'... and there is a *lot*. Indeed, data overload is a major feature of this technology. But, equally, precisely because access is fast and simple, one need not even make a trip to the library. The 'net has hastened a 'reseach' society (bemoaned by Heidegger since the library is less and less used...)

- Second, what is there is often non-selective. Since anyone can get something onto the 'net, it is not even like a library which must decide what to subscribe to and what not. Thus the user of the 'net must provide his or her own selectivity. Jewels and junk are all there. One must develop a 'net 'savvy' to sort out the good from the bad.

- I have already previously mentioned the scams (now more sophisticated in, for example, mimicking banks: you must re-verify your account numbers?); spaming; pop-ups; junk mail; viruses; and the hundreds of kinds of irritations which constitute minor stupidity. Now, I do have to admit that I fall into an intergenerational pattern here—I deeply dislike this 'noise' since it slows me down in relation to the tasks I want to perform, whereas my 22 year old son tells me that the very 'chaos' of the net is likeable and part of its excitement.

- Then there are the versions of 'avatars' and deceptions which occur—in the US, the 'wars' of pedophiles and child pornographers searching for prey, chased by detectives and investigators skilled at mimicking pubescents continues. Our Vermont soccer coach was caught in just such a sting operation.

I need not go on to list the minor, and sometimes major stupidities which occur in the mediations which constitute the knowledge society. But I would rather now turn to some interesting examples of possibilities and experiments which are also part of this style of global, electronically mediated worlds:

- Last year I reviewed Neil Gershenfeld's latest book , *FAB :The coming revolution on your desktop—from personal computers to personal fabrication*. See, yet another modernist 'revolution', this one in often utopian form. [Gershenfeld is the past director of MIT's media lab, but now works on what he calls a FAB lab, a fabrication lab which uses three-dimensional, material 'printing' to produce things.] Part of his strategy is to distribute relatively inexpensive 'labs' into the developing world and to encourage users to come up with their own uses—he wants to invent inventors. The example which struck me was that of Sugata Mitra, an Indian computer scientist. His office was next to a major slum in Delhi, with the back wall of his office a sort of barrier to the slum area itself. "As an experiment, he opened a hole in the wall and stuck into it an Internet-connected computer facing the alleyway. There was no explanation, no instruction, just a computer monitor and a joystick [mouse]."[11] Within minutes kids arrived and soon learned how to surf the 'net; they went on to organize their own 'classes' and to pass on information to their friends, many could not speak a word of English. "Because they were self-taught, the kids developed their own explanations for what they were doing. The mouse was called a "sui", the Hindi word for needle....and the hourglass icon was called a "damaru," after the shape of Shiva's drum."[12] And they developed their own skills at use. Then, when Sugata came back one day, he discovered the words, "I love India" on the screen and since there was no keyboard, he was mystified about how the kids had accomplished this. "They explained hat they had discovered the 'character map' program on the control panel that can be used to enter arbitrary characters with the mouse by clicking on an on-screen keyboard. Sugata has a Ph.D. and he didn't know how to do that."[13] Unintended use and consequence for sure. This experiment was repeated with similar results in many locations.

This, clearly, is *not* an instance of stupidity. Yet, in another sense, it also *does not* fit the Drucker notion of knowledge work or the knowledge society. The kids, after all, were neither specialists, nor formally educated, yet they *innovated* a way of doing something even the expert who set up the experiment did not 'intend' or 'know.' This inspiring event is the inverse, however, of one of my earlier obscure vignettes. If WMDs [weapons of mass destruction] are not available; then WSDs [weapons of simple destruction] will do. What we are witnessing here is the *multi-*

---

[11] Neil Gershenfeld, *FAB: the coming revolution of your desktop—from personal computer to personal fabrication* (Basic Books, 2004), p. 203.
[12] *ibid.*, p. 204.
[13] *ibid.*, p. 204.

stable role all technologies can play. The mobile phone, with its multitasks, can also serve to detonate the WSD, just as my laser pointer may disable a video surveillance camera. It is precisely for these reasons that technologies are not and cannot be simply *deterministic,* and for that reason not simply predictable, rather, they are *overdetermined* as these examples illustrate. I am, for example, willing to wager that the determination of the Chinese government to restrict and censor the 'net will fail, at least it will fail for users equivalent to the Delhi kids who find a way to get around a lack of a keyboard in a hole-in-the-wall computer. Why do I suspect that Google, too, knows that? What we are seeing here is a reason why technology predictions so often fail. Within the human-technology interaction, there is overdetermined multistability of a very complex sort.

This leaves me with but one more of my opening irrelevant vignettes, that of the Danish cartoonists who did not bomb the Samarra Golden Mosque. This vignette is the most sensitive and perhaps the hardest to deal with, given the current world tensions. Instead, take an example from another culture. The recent discovery and publication of a *Gospel of Judas* portrays Judas as the secretly best friend of Jesus, who at Jesus' request turns Jesus over to the authorities in order to fulfill his destiny. A *New Yorker* cartoon capitalizes on this with a cartoon portraying Jesus and Judas in a bar, happily singing and enjoyed each other—surely likely to be offensive to religionists.

Similarly, one can find at least one form of resistance to the popularity of alternative origins stories. For example, spokesmen from the Vatican are urging a boycott of the movie, "The DaVinci Code." But no movie theatre is likely to be trashed. But, again, what does this have to do with the knowledge society and its technologies? This example is somewhat complex in that it shows something about how the seemingly 'same' technologies are differently embedded in *different* cultures. We are all aware of the response to the Danish cartoons in Islamic countries and cultural elements of the clash are also all too obvious. I shall not venture to guess where across the spectrum of opinion the tolerance for satire and irony ceases—Salman Rushdie, after all, still carries part of the Islamic *fatwa* against him as well. And although when the Danish cartoons were disseminated in the Middle East, the situation was complicated and exacerbated by the political actions of the Mullahs who included even worse cartoons than those published, to make the situation even more incendiary. The subesquent public demonstrations and violence clearly were larger than one would have expected and showed that the line across which one might accept irony and satire, is clearly narrower than in modern culture.

Knowledge society technologies play a role in all this: the 'net, television, paper publications, DVDs are all part of the instantaneous dissemination network. Interestingly, these same technologies are being used to disseminate the threats against the West, the new "crusaders," –deliberate terminology designed to conjure the *Medievalist* tone extant in Islamic fundamentalism—are conveyed by audio-video means. Osama bin Laden and al-Zarqawi use audio and audio-video tapes, then transmitted to El Jazeera television to equally speed the messages. If 'modernism' is to be rejected; it is rejected only in part since both the communication technologies and the weaponry are thoroughly modern. Islamic fundamentalism is *not* like the Amish, who carefully consider just which technologies can be adopted into their culture by considering first which ones enhance and which ones distract from their communitarian values.

Let us now flip the coin to its other side. I hold that all technologies are *non-neutral*, they carry with them selectivities which amplify certain, and reduce other possibilities in use. It has been of interest to me lately, that in spite of attempted program control (except for fairly wide spread satellite television) that there is growth today in two media in Islamic countries: television news in particular, and cinema. Our PBS [public broadcasting system] television did a documentary on the growth of news program competition in the wealthy Middle Eastern countries [Dubai, Saudi Arabia, Kuwait, etc.]. While El Jazeera retains the largest audience, special interest groups have whittled away at the overall audience. This seems like a normal modernist technological development—but it is not what I noticed. Instead, I noticed that attractive young women were *included* amongst the reporters and announcers, and even served as some anchors; these young women were clearly well educated; and, most important, *not a single one wore a burka or chadour, but in fact, none even wore head scarfs!* Here the news television 'culture' if one can call it that, trumps the local culture. I have previously argued that modern media technologies are *acidic* to traditional cultures, and this is a good illustration of that point.

Regarding cinema, a different point may be made. The *New York Times* printed a review of a new movie made in Saudi Arabia.

1. Stupidity in the Knowledge Society

Saudi Arabia has very few movie theatres, of those extant, many are subject to closure and censorship—yet the new, often daring, film makers are featuring young people, including again, scarfless women. And, for avid movie goers, I suspect many of you are already familiar with a good number of excellent Iranian movies: "Kandahar" is one of my favorites, but there are many more. I am not suggesting that images of bikini clad young women, or acceptance of ironic or sarcastic cartoons are about to be accepted yet, but I am willing to wager that cinema will make serious inroads into traditional cultures. Technology transfer is also culture transfer and with cinema technologies, the transfer becomes easier with miniaturization, DVDs, and small portable players. The very pirating of intellectual property, another product of the knowledge society, makes it easier still.

I have now re-connected with each of my opening vignettes, placing these in the context of the technologies of the knowledge culture. There remains only one more debt: the much deeper and difficult problem forefronted by *Scientific American's* crossroads—what is to be done about our rapid use of the planet's natural resources? This issue overshadows all the issues I have mentioned in spite of the fact that it has not been and cannot be the primary focus of this occasion. So, I can only look at one tiny piece of our dilemma, and I hope what I am about to say is not taken as pandering to my Danish audience, but as a genuine personal vignette.

This is approximately my 25th trip to Denmark in the last dozen years, and it seems that on every trip, as my plane gets onto its glide path, that I see *more Danish windmills*. I like Danish windmills—I find them simple and with their slow turning blades—look like a sort of elegant technological choreographed dance show. And, yes, I am aware of the national policy; of how it was instituted; and how it is working. And, I agree, that such renewable resources are part, however small, but significant, to the need for energy and the escape from dependence upon fossil fuels. I do not know if Heidegger would have like them or not, since they are behemoths; they can be linked into power grids; and all the 'bad' things he associated with modern technologies. But they do move only when the wind blows rather than transforming 'nature' into standing reserve.

And, I also know that GE (General Electric) is now building even bigger ones and these are beginning to play into the scene in both my own locales. I live on Long Island, and windmill farms are being proposed out at sea off the south shore; similarly, my vacation home is in the mountains of Vermont and there is a proposal to line the mountaintop of Magic Mountain, now a ski area, with a row of GE windmills. [There is a more publicized project being proposed off Cape Cod, near the Kennedy vacation area in Hyannisport.] All these proposals are being furiously opposed! The primary reasons given are 'aesthetic'—it is claimed that such large windmills look too 'industrial.' But fear of bird kills, disruption of navigation, fishing obstruction, and a slew of other reasons are also claimed. Yet, the secret lies in who the players are: In the case of Hyannisport, one of the big opponents is a politician from Alaska, long linked to oil corporations and long committed to drilling in the Alaskan wilderness. He is funding a lot of the opposition to the Cape Cod project. In the case of Long Island and Vermont, wealthy persons who could see these windmills, again those who have, in many cases, made their fortunes from large corporations, and those who do not in any way need local employment are the most vociferous opponents. The 'locals' in Vermont seem much more positive, almost in an inversion of early industrial times where those with cottage industries were the ones threatened by factories.

Of course the American scene is very different from the Danish one. Our windmills are being pushed by mega-utilities with little case being made for local benefit, so windmills in my country play a different role than here. The 'same' technology, once again, is differently embedded and I do not know now which direction will be taken. For my part, even as a land owner, who will see them in the winter, I like windmills.

2

The Designer Fallacy and Technological Imagination

Abstract: Most literary critics have abandoned the notion that the meaning of a text lies in the intention of the author and have called this the "intentional fallacy." I hold that there is a parallel found in many interpretations of technology design and call it the "designer fallacy." This chapter, through examining a wide series of historical technology designs, deconstructs the utility of a simple designer-plastic material-ultimate use model and suggest that one must take into account unintended uses and consequences, the constraints and potentials of materiality, and cultural contexts which often are complex and multistable. I outline a complex, interactive account of design interpretation.

Key words: Designer intent, designer fallacy, complex and multistable relations, technofantasy.

Earlier in the 20th century, literary theorists developed the notion of an "intentional fallacy." This was the notion that the meaning of a text lay with the author's intentions – if these could be uncovered, then the meaning of the text was established. One can easily see how, if this is the only true way to establish meaning, there could be difficulties. What if the author was long dead? Or, even if living, how could one tell that the author was himself or herself telling the truth? What of unintended meanings, or meanings which fit but were not thought of in advance? Thus, the intentional fallacy recognizes such difficulties and cannot be considered an adequate account of interpretation.

I hold that there is a parallel 'fallacy' which is at least implicit in the history of technology design. In simple form, the "designer fallacy," as I shall call it, is the notion that a designer can design into a technology, its purposes and uses. In turn, this fallacy implies some degree of material neutrality or plasticity in the object, over which the designer has control. In short, the designer fallacy is 'deistic' in its 18th century sense, that the designer-god, working with plastic material, creates a machine or artifact which seems 'intelligent' by design – and performs in its designed way. Instead, I hold, the design process operates in very different ways, ways which imply a much more complex set of inter-relations between any designer, the materials which make the technology possible, and the uses to which any technologies may be put. Ultimately I am after a deconstruction of the individualistic notion of design which permeates both the literary and technological versions of the fallacy. First, some examples of simple designer fallacies: Thomas Edison, the great late 19th-early 20th century American inventor, was among the first to design and invent a machine to reproduce sounds – the phonograph. The machine, at first, was a mechanical device which consisted of a speaking tube into which someone would speak; this was attached to a sensitive diaphragm which would reverberate with the sound waves coming into the tube and the diaphragm, in turn, was connected to a crystal needle which would trace the wave patterns onto a rotating roll covered with tinfoil. As the crank was turned, the speaker sounding into the tube, a 'record' was made on the foil. The same machine, played back, would reverse the process and one could hear, well enough to understand and recognize the sounds, originally inscribed on the roller – "Mary had a little lamb...." (Nyre, 2003)

Here, the designer intent was to reproduce sounds. But the intent, at this stage, remained ambiguous and the primary possible use of this machine was drawn from the resultant capacities which emerged, more than from any pre-planned single use. It could be a rather primitive dictation machine. Clearly, it would have restricted use since the number of play-backs was very limited due to the softness of the foil – the play-back would remain intelligible for only one or two times. In spite of this, the machine was advertised in the typically glowing rhetoric of technological promise of the late 19th century. It was advertised as "The miracle of the 19th Century," a machine that speaks:

> It will Talk, Sing, Laugh, Crow, Whistle, Repeat cornet solos, imitating the Human Voice, enunciating and pronouncing every word perfectly, IN EVERY KNOWN LANGUAGE." (Nyre, 2003)

If one, with the anachronistic insight of knowing anything about the subsequent history of recordings, read back to Edison's early machines, one might have predicted that one early dominant use

2. The Designer Fallacy and Technological Imagination   21

of recording devices would quickly evolve into music recording, which in turn, also transformed a number of musical practices. For example, early recording devices could record for only three and a half to four minutes of time—thus the music played must be three and a half to four minutes long, a traditional length for the 'popular song' which persisted well past the time of early recording devices. The new machine calls for new practices, but in this case not 'intended' ones.

The phonograph came later than the telephone, invented at least once by Alexander Graham Bell. Here the designer intent was for an amplifying device capable of transmitting a voice over distance, and intended as a prosthetic technology for the hard-of-hearing (Bell's mother). The early antecedent of "chat" on the internet, the party line on which all the neighbors 'chatted' was not foreseen, let alone the subsequent telephone wiring of early 20th century America.

Even the typewriter was first designed as a prosthetic technology aiding blind or myopic people by allowing them to produce clear script. Instead, as Friedrich Kittler has pointed out, the typewriter became, dominantly, a business machine and one which transformed the secretary of the late 19th century from male to female (male secretaries often refused to adopt to this 'machine' which they thought deskilled their handwork, but young women, seeking both a public role and pre-skilled with keyboard or piano skills, easily found a new role)! (Kittler, 1990) The designer fallacy also plays a role in Langdon Winner's best-known story, "Do artifacts have politics?" This article traces the history of Robert Moses' designs for the bridges over the parkways of Long Island. Winner claims that Moses' ulterior intention was to keep the lower classes and races out of Long Island's pristine growing suburbs. Thus he deliberately designed low bridges which would prevent large trucks and double decker buses from using the parkways. In one sense, there was some success with this material strategy if one looks at the demographics of the early 20th century – but a counter-strategy defeated whatever politics were first employed. The Eisenhower Interstate development of the 50s called for all interstate highways to have high bridges so that trucks – including those carrying ballistic missiles for the Cold War – could clear them, thus opening the way for what we Long Islanders call our "longest parking lots" of multi-laned highways. The Cold War

2. The Designer Fallacy and Technological Imagination

trumps suburban protection.[1]

The language and notion of 'intent,' while still dominant, is inverted by Edward Tenner whose well known book, *Why Things Bite Back: Technology and the Revenge of Unintended Consequences* (Knopf, 1996). Tenner catalogues and classifies an enormous number of technologies, presumably designed for certain uses, which end up having disastrous or contrary unintended consequences. He spoofs Toffler's notion of the paperless society, where, "making paper copies of anything is a primitive use of [electronic word processing] machines and violates their very spirit, in light of the higher-papered society of today. (Tenner, 1996). Or, something as simple as a home security system, designed to increase security, he contends subverts security by producing false alarms and overwhelming police ability to respond, "In Philadelphia, on 3,000 of 157,000 calls from automatic security systems over three years were real; by diverting the full-time equivalent of fifty-eight police officers for useless calls, the systems may have promoted crime elsewhere." (Tenner, 1996) Tenner's examples are of *unintended,* but also of *unpredictable* effects. The patterns being traced here apply equally to simple and complex technologies. I have lived through the long term claim of virtually infinitely free energy to be produced from nuclear sources, through the Three Mile Island near melt-down situation, to the closing of Long Island's Shoreham nuclear plant, designed as part of this trajectory of designer intent, but which to date has ended in a colossal, $4,000,000,000 'technology museum' which as yet has no use.

From the comparatively simple examples above, one can note that designer intent may be subverted, become a minor use, or not result in uses in line with intended ends at all. In addition, with unintended consequences the theme becomes the unpredictability of the uses of technologies. But, there remains a persistence of the designer fallacy, that in some way 'intent' determines, however successfully or unsuccessfully, outcomes. My argument is directed *against* this framing and description of the design project. What I hope to establish is a description which recognizes much more complex relations between designers, technologies and the ultimate uses of technologies in variable social and cultural situations. My approach is descriptivist in a sense parallel to those in

---

[1] In discussion, it was pointed out that there is a difference between initial design intent, and subsequent design modification, but the argument I am making is that in neither case is there simple designer control over outcomes.

science studies and the history of science which eschew end results over the examination of development in process (Kuhn, Latour, Pickering). I will open the way to my counter-thesis by looking at several variations upon technologies and the embedded ways in which these function. Again, I am arguing against an individualistic notion of design, and for a more complex set of relations between multiple inputs into developing technologies and for multiple, multistable possibilities for any single technology.

First, I want to show something of how technologies are differently _embedded_ in different cultural contexts. My first example is the windmill—a device which like a pinwheel turns with the wind. The most ancient example, according to Lynn White, Jr., is to be found in India, a wind-driven prayer wheel or 'automated praying device.' (White, Jr., 1971) There were, and continue to be, hand-driven prayer wheels, rotating drums on a hand-held handle, which can have written prayers on the surfaces which are then spun with the prayers presumably being sent outwards. The 'automated' prayer wheel of the wind driven device lets 'nature' do the work. Later, in Mesopotamia, larger versions of the windmill occurred in the 9th century. These devices were used to provide power for such applications as milling. Moving to Europe, 'windmill fields' were developed to help pump out the lowlands of Holland in the 9th century in an early 'technological revolution' of larger-scale power use. Finally, today, we are moving into the argumentative phase of wind-generated energy, well accepted and in place in Denmark, which produces nearly 20% of its energy from windmill farms. In England and the USA, such windmill farms, proposed for offshore or mountain ridge sites, are undergoing technology assessment battles along NIMBY [not-in-my-backyard] lines.

Abstractly, one can argue that these are all the 'same' technology, wind driven devices to supply different powers, but each example is differently culturally embedded. The need to have relatively constant praying is quite different from the need to have renewable energy, and to call each a different 'use' is to abstract from the complexity of the cultural background. The 'same' technology is embedded differently in the different historical-cultural settings. But this is also to say that the 'same' technology can fit into different contexts and is _field located_.

A closer look, however, also shows that what I have called the 'same' technology, is also materially different in each context. The Indian wind-driven prayer wheel is a relatively small device, whereas the Danish and contemporary high-tech windmill is up to a 100 meters tall; and the former responds to the speed of the wind with faster or slower revolutions, whereas the latter turns at the same speed through self-governing blade adjustment. Both entail what Andrew Pickering calls a process of "tuning" and a "dance of agency" in the development process. (Pickering, 1995)

In design, the "tuning" and "dance of agency" can often turn around 'designer intent.' Bruno Latour has made the familiar post-it example famous in *Science in Action* The designer, experimenting with the material properties of various glues, accidentally as it were, produced a glue which would stick only temporarily—thus seemingly a failure in terms of 'designed glues.' But, instead of simply casting aside the new propertied invention, the designer began to think of possible new uses and chanced upon the idea of page marks for hymn books. (Latour, 1987) Thus, a new use, both unintended and unplanned, led to what today is a massive market for post-it products. One could say, were one to adopt Latourean language, that the non-human here transformed the human (designer) with its actant, material behavior! I have frequently employed a similar example. Take the million year old 'hand-axe,' the chipped tool from pre-modern hominids which is usually thought to be a scraper and butchering tool, although no one knows the possible uses which could be many, and the small, sharp earlier-thought-to-be-detrius chips from the hand-axe, which are now recognized to have been used for cutting and even, possibly, surgery, and we get an archaic version of the post-it story.

Allow a quick pause with respect to the designer-intent model of technological development: it should appear by now that the 'designer fallacy' may well be the rule rather than the exception. While it may be the case that some technologies have come into being and performed as 'intended' by their designers (I admit, I can think of none which have served solely in this way), there would seem to be none which can not be subverted to other, to unintended, or unsuspected uses and results. This is frequently the case for an initial design and even more so for later modified designs. Moreover, whether simple or complex, the same indeterminacy seems to apply. As artifactual, technologies seem potentially to contain *multiple uses or trajectories of development.* If even the simplest artifact, an Acheulean hand axe, can be used for multiple purposes, it differs little in outcome from the purposely designed multi-task tool, the Swiss Army Knife. Indeed, multi-tasking may be an emergent pattern for contemporary technologies. Some have begun to hold that the trajectory of multi-tasking for information technologies, is toward a single big and a single small multi-tasking instrument. The mobile technology which, like the Swiss Army Knife, is a cell phone, digital camera, bar-code reader, email device, etc., etc. is the single small multi-tasking technology, while the large home entertainment unit (TV, DVD, computer screen, etc., etc.) connected to the economic, entertainment, communications dimensions of life, is the big multi-tasking instrument; and while much of this remains technofantasy, it is plausible technofantasy.

Fantasy, however, is one type of *imagination* which also plays a role in, behind and throughout design activity. I think a case can be made that in the high Middle Ages, a form of technofantasy began to emerge which, at first slowly, but with acceleration, began to shape the form of culture in Europe, which in turn pointed towards the saturated technological culture of today. Lynn White, Jr. has argued that there was something of a technological revolution which occurred in this period. The construction of high-standing Gothic cathedrals called for machines and architectural techniques not employed previously. Admittedly borrowing interculturally from, first the Moorish styles which entered Europe no later than the 10th century, but taking these to greater extremes, Chartres, Notre Dame, Cologne, all borrowed flying buttresses and glass-stone frillery. What might not be noted, however, was a similar shift in imagery in the world of fantasy. The fantasy paintings of the Bruegels remained largely 'organic' or 'animal-like' fantasies. Devils, dragons, demons, large monsters, clearly were 'biomorphic' however fantastical. But by the 13th century, machines began to play fantasy roles. Roger Bacon described fantasy machines, such as self-propelled ships, underwater craft, flying machines and other impossible-to-build machines for the times, machines which were later 'visualized' in the 15th century by da Vinci in his notebooks (discovered and publicized by the Futurists in the 1920s). I am hinting that a specific mode of technology imagination or fantasy began to take hold. This probably was a life-world reflection, since many of the radical new machines which began to appear and be developed in Europe had earlier, in other forms, come from the multicultural trade, journeys and experiences of the cross-cultural exchanges between Islamic culture, the Mongolian invasions, and the post-Marco Polo adventures to the Far East. Lynn White, Jr., Joseph Needham, and others began to recognize this cross-cultural trade of technologies by the middle of the 20th century. Spices, gunpowder, the compass, silk, windmills, as previously mentioned, all migrated to Medieval Europe, and were adapted and developed. Optics, better known by Al Hazen (1038) than the West, ended up on a trajectory of lens making which led to the optical inventions of the telescope and microscope which drove the early scientific revolution, instrumental technologies provided the infrastructure of science itself.

All of this today is relatively common tender. But it needs to be seen in the light of the 'designer fallacy' I am addressing here. Each new invention which came into Europe, often first a matter of fascination, became adapted into new uses and developments. While China invented gunpowder, it did not successfully produce a *cannon!* But by the Thirty Years War, cannons were being used to demolish French castles at the rate of dozens per week. (DeLanda, 1991) It is with this observation that I will now begin my move away from the 'designer fallacy.'

However some material capacity comes to human awareness (discovered by accident, through experiment, through found discovery, or – I suspect rarely – planned out from design) once that capacity is emergent and clear, some possible 'trajectory' is suggested. One could say, the explosiveness of gunpowder "suggests" uses. But, those uses will also be likely to be culture-relative, at least at first. Long before the cannon, feudalism had produced the land-castle system, wherein the lords who were to protect the populace had built defensive keeps. A many centuries-long form of contest centered on strategies of defense with supplies and means of defending against the attackers, a strategy which tended for a time to favor the well stocked and designed castle. Siege machinery, too, grew in complexity over the centuries, in an evolution from Roman times with trebuchets, catapults and the like. None of these engines, however, could easily breach walls – which the cannon could do.

In terms of design history, the cannon is in a sense pre-modern. No one knows who 'invented' the cannon, although many attempts to create a workable cannon were made, including the production of early, fire hardened wooden cannon barrels (not too successful). The cult of the individual designer had not yet come into being. Visiting Meissen in Germany recently, my guide, Professor Bernhardt Irrgang, pointed out that the cathedral there had a room for the architects, and while names of leading architects were sometimes known, the name actually served as something

2. The Designer Fallacy and Technological Imagination 27

of a 'school' of such and such an architect – the same was often true for Renaissance artists. The room or office was for the whole entourage assigned the task of keeping the cathedral in repair. As Foucault has pointed out, the same frequently applied to authors – individual authors came into being with modernity, thus pointing to an even deeper connection between the "intentional" and the "designer" fallacies.

Let us now return to the designer problem and begin re-casting it. I wish to focus upon two interstices in a three part relation. The first interstice, in simplest form, is that between the designer-inventor, or including subsequent designers and materiality. What is at play is a set of interactions between the designer(s) and the materials being worked with – it is a two-way relationship within which the "accommodations" and "resistances" Pickering speaks of, come into play. (Pickering, 1995)

My beginning example is the long fantasized desire of humans for flight. The Icarus story, with its technologies of bird feathers and wax, is clearly fantasy only. Similarly, Roger Bacon's and later Leonardo da Vinci's descriptions of flying machines also remain in the imaginary realm, although da Vinci's recognition of the curved wing shape of birds was a step in the right direction. Almost everyone has seen documentaries on early flight experiments, usually comic with films of flying contraptions – human powered – and their subsequent falls and crashes. But, note, once again, the serious experimentation begins with that Industrial Century, the 19th.

From the beginning, it was recognized that wings had to be both light and strong, and the design was at first biomorphic in that bird wings, and sometimes batwings, served as the pattern. Yet, how clumsy the designs seem in retrospect! Gliders began to succeed to some degree, with much experimentation of light materials, wood or bamboo, and glued linen or other light cloth. Interestingly, the reluctance to follow the fantasy trajectory of human powered flight gave way to the recognition of the need for a light-weight power source which historically we recognize as the internal combustion engine plus 'screw' or propeller. The Wright brothers' flying machine was a hybrid conglomeration of many technologies. The Wright brothers were experienced light weight technologists – bicycle makers – who adapted from windmill technologies, a propeller for driving through air rather than being driven by the wind. Then making wing and control designs, some modified from other's attempts, they eventually produced the first powered flights (I ignore the historical controversies around who actually first flew, since there were many contenders). What we really have in this history is a competitive 'dance of agency' through trials and failures, until finally the small success which launched the trajectory of human+machine flight. From 1903 to the present century, development has seen flight move away from biomorphic designs towards ever more variations of flight which are less and less like those of flight's origins. The simplest example is that of a fixed wing over a flexible and moveable wing. Flight, originally fantasized as embodied human flight, has never really materialized, its closest actualization probably is that of hang-gliding and its kin, which flight is restricted to lovers of extreme sports. The one bicycle-technology, propeller driven, light-weight aircraft, flown by a trained cyclist, which successfully flew across the English Channel, was hardly anything like birdlike grace in form, even if actually human powered. But with mylar skin, and weighing in at only pounds, it was a culmination of a trajectory towards lightness which was the material need for this approximation of flying. What I am trying to point out, is that one does not find anything like sheer plasticity of the material, over which the designer has anything like a transparency of control. Rather, one finds a process of interrogation of materiality and experimentation with it, which results – sometimes – in fortunate results.

The second interstice would, under the designer fallacy model, be the 'uses' to which the invention, the technology, is put. Maintaining the analogy to literary practices, this would be reader response, or responses. What results from the literary or technological product? In the case of my flight example, the proliferation of uses is historically clear – there is something like an actualization of a possibility tree. In less than a decade, airplanes were beginning to be used militarily, by World War I, there were inter-airplane "dog fights", bomb dropping, reconnaissance; equally early, commercial developments began; recreational uses with the "barnstormers" and stunt fliers; races, distance breaking flights such as Lindberg's over the Atlantic and the like. And, in each use, changes in previous practices occurred. By World War II, the *Blitzkreig* employed its own version of "Shock and Awe" with Stuka dive bombers, to the present, where unmanned Predators and 'smart bombs' are employed, displacing what was once trench warfare or disciplined regiments marching at one another. I need not follow each of these trajectories, but it is clear that Orville and Wilbur neither foresaw the speed or the diversity of

2. The Designer Fallacy and Technological Imagination 29

their invention's results. And, just as the interrelation of designer and materiality contains an indeterminate set of accommodations and resistances, through which may be produced a result never simply planned, so with the results and the indeterminacy of multiple uses.

I have tried to show that the designer-materiality interstice is such that the inter-relation of designer-materiality precludes any simple notion of control or transparency over the simply plastic or passivity of the material. Instead, the interaction is exploratory, and interactive. In the second, now artifactual-use interstice, the designer has even less control or impact, rather the user(s) now play the more important role. The indeterminacy here is multistable in terms of the possible range of uses fantasized or actualized. One particular set of interesting examples comes from the ingenious ways in which technologies may be defeated – defeasibility uses. Video surveillance cameras, for example, may be disabled by laser pointers flashed into the lenses. Hardened steel steering wheel anti-theft devices, precisely because hardened steel is vulnerable to fast-freeze brittleness, can easily be broken when sprayed with a freeze spray. Slightly more complex are the 'wars' between police determined to trap speeders with radar, now laser speed detection devices and the 'insurgencies' which develop technologies to detect radar signals or confuse laser reading devices. And so go the multiple directions from same, different, or differently used technologies.

We are now in a position to draw a few conclusions from this examination of designer fallacies. First, in spite of language concerning designer capacity in textbooks – recognizably there in engineering, architecture and other design textbooks – I am attempting to show that the design situation is considerably more complex and less transparent than it is usually taken to be. Both the designer-materiality relation, and the artifact-user relations are complex and multistable. While it is clear that a new technology, when put to use, produces changes in practices – all of the examples show that – these practices are not of any simple 'deterministic' pattern. The results are indeterminate but definite, but also multiple and diverse. Moreover, *both* intended results and unintended results are unpredictable in any simple way, and yet results are produced. And, finally, what emerges from this examination looks much more like an inter-relational interpretation of a human-technology-uses model in which the human, material, and practices all undergo dynamic changes. If this is the case, then there are also implications for designer education. One of these is that the design process must be seen to be fallibilistic and contingent. Some worry that this recognition may be demotivating – but it could also be a call for a more cooperative, mutually co-critical approach as well.

I am also implicitly suggesting that the re-descriptions which have arisen out of the past several decades of work in the history and philosophy of science, the new sociologies of science, and cultural and science studies, which undertake careful case studies of developments in technologies, give hints of the complexities suggested.

References

DeLanda, M., 1991, *War in the Age of Intelligent Machines*, Swerve Editions, Zone Press, New York, pp. 12–14.

Latour, B., 1987, *Science in Action*, Harvard University Press, Cambridge, MA, p. 140.

Kittler, F., 1990, The mechanized philosopher, in: *Looking after Nietzsche*, Laurence A. Rickels, ed., SUNY Press, Albany, NY.

Nyre, L., 2003, *Fidelity Matters: Sound Media and Realism in the 20th Century*, Doctoral Dissertation, Department of Media Studies, University of Bergen, Volda University College, Norway, pp. 89–90.

Pickering, A., 1995, *The Mangle of Practice: Time, Agency, and Science*, University of Chicago Press, Chicago, p. 102.

Tenner, E., 1996, *Why things Bite Back: Technology and the Revenge of Unintended Consequences*, Alfred Knopf, New York, p. ix.

White, Jr., L., 1971, Cultural climates and technological advance in the Middle Ages, *Viator* **2:** p. 139.

3
Aging: I don't want to be a Cyborg

Abstract: Examination is made of a range of cyborg solutions to bodily problems due to damage, but particularly related to aging. Both technological and animal implants, transplants and prosthetic devices are phenomenologically analyzed. The resultant trade-off phenomena are compared to popular culture technofantasies and desires and finally to human attitudes toward mortality and contingency. The same resistance to contingent existence and becoming a cyborg is noted.

Key words: cyborg technologies, desire, implants, prostheses, prosthetic devices, technofantasy, trade-off, transparency, transplants.

Although the term, cyborg, was not invented by Donna Haraway, it was she who helped make it a term of popular culture. Her cyborg is a *hybrid* which can include human, animal and machinic or technological parts. These, in turn, may combine with various human fantasies, including *technofantasies*. In popular culture, these fantasies now include utopic-bionic science-fiction variants which in filmic and televisual form include all sorts of prostheses which fantasize making mere human limbs, organs and the like super-powerful, and better than the originals. *"Terminator," "Robo-Cop," "Bionic Man/ or Woman,"* have all had their plays upon this theme. And, each is a variant upon what I claimed was a deep rooted technofantasy desire which I described in *Technology and the Lifeworld* (1990):

> There is a deeper desire which can arise from the experience of embodiment relations. It is the doubled desire that, on the one side, is a wish for total transparency, total embodiment, for the technology to truly "become me." Were this possible, it would be equivalent to there being no technology, for total transparency would be my body and senses; I desire the face-to-face that I would experience without

the technology. But that is only one side of the desire. The other side is the desire to have the power, the transformation that the technology makes available. Only be using the technology is my bodily power enhanced and magnified by speed, throught distance, or by any of he other ways in which technologies change my capacities. These capacities are always different from my naked capacities. The desire is, at best, contradictory. I want the transformation that the technology allows, but I want it in such a way that I am basically unaware of its presence. I want it in such a way that it becomes me. Such a desire secretly rejects what technologies are and overlooks the transformational effects which are necessarily tied to human-technology relations.[1]

Now, in that earlier context I was describing what I have called *embodiment relations,* which were experienced uses of technologies which remain detachable, but which in use are quasi-transparent, and not technologies literally taken into or inside my body. Yet, I will maintain that the desire remains applicable to *cyborg technologies.* To make the case, I will need to account for several historical as well as contemporary variations upon technologies used both in detachable and non-detachable, internalized forms.

One of the oldest such anticipations of cyborgization are *prosthetic devices.* False teeth, peg-legs, arm hooks, and various devices to replace lost teeth, limbs and such are very ancient. These prosthetic devices, substitutes for lost body parts, remain detachable and thus fall under the earlier descriptions I have made concerning embodiment relations—one experiences one's surroundings through the quasi-transparency of such devices, but always with a detectable difference which magnifies some and reduces other features of one's experienced environment. The peg-leg can never 'feel' the hot, sun-baked sidewalk which the bare foot would feel—but slid along the rough texture, one might even better than with toes, 'feel' the rough textural features of the surface. Nor would I expect that the users of such ancient devices could easily fall into the slippery slope and utopian fantasies which so frequently dominate our science fiction and virtual reality hype contexts and which describe these devices as 'better' than the lost body part—or, am I wrong? Pinocchio, after all, capitalizes upon a fantasy of a dummy-become-alive! The desire remains, but the

---

[1] Don Ihde, *Technology and the Lifeworld: From Garden to Earth* (Bloomington: Indiana University Press, 1990), p. 75.

device remains "dumb." No one in a right mind would likely seek an amputation which would replace one's healthy limb with a wooden one? But most people would choose to have a prosthesis once the limb is gone in order to restore some semblance of motility and capacity. The proto-cyborg is thus a compromise.

Vivian Sobchack, a phenomenologically trained scholar with a hi-tech prosthesis brings us up to date in a specific response to the quotation above:

> Obviously, transparency is what I wish—and strive—for in relation to my prosthetic leg. I want to embody it subjectively. I do not want to regard it as and object to think about it as I use it to walk. Indeed, in learning to use the prosthesis, I found that looking objectively at my leg in the mirror as an exteriorized thing—a piece of technology—to be thought about and manipulated did not help me to improve my balance and gait so much as did subjectively feeling through all of my body the weight and rhythm of the leg in a gestalt of intentional motor activity... So, of course, I want the leg to become totally transparent. However the desired transparency here involves my incorporation of the prosthetic—and not the prosthetic's incorporation of me.... This is to say that although my enabling technology is made of titanium and fiberglass, I do not really or literally perceive myself as a hard body—even after a good workout at the gym, when, in fact, it is my union with weight machines (not my prosthetic leg) that momentarily reifies that metaphor.[2]

This is a good description of, precisely, an embodiment relation, of *quasi*-transparency. Sobchack, however, is not tempted by the slippery slope utopic slide:

> Nor do I think that because my prosthetic will, in all likelihood, outlast me, it confers on me invincibility or immortality. Technologically enabled in the most intimate way, I am, nonetheless, not a cyborg. Unlike Baudrillard, I have not forgotten the limitations and finitude and naked capacities of my flesh—nor, more important, do I desire to

---

[2] Vivian Sobchack, *Carnal Thoughts: Embodiment and Moving Image Culture* (Berkeley and Los Angles: University of California Press, 2004), p. 172.

34   3. Aging: I don't want to be a Cyborg

disavow or escape them.³

Even more hi-tech prostheses built with springs for below-the-knee amputations have allowed highly motivated and skilled athletic persons to actually achieve high running speeds, such as those demonstrated by Jami Goldman, amputee sprinter. Perhaps the most famous model-athlete is Aimee Mullins who, with her spring-legs, played Cheetah Woman in Matthew Barney's "Cremaster 3" movie series. Mullins turns in highly respectable records running, but also alternates her spring legs with other prosthetic legs, include a pair of glass limbs, for other purposes. She has had prostheses since learning to walk because she was born with fibular hemimelia (born without fibula bones) and underwent amputation at age one. Growing up with prostheses is probably as close as one can get to minimal quasi-transparency. Prostheses with spring components have today become more common with the large numbers of Iraq War amputees as well, yet all remain within the noted degrees of limitation cited for detachable devices.

Before leaving limb prostheses, I should mention their internal and permanent counterparts—knee, hip joint and other implants substituting for bone and cartilage damage. Stainless steel and Teflon, restore to a degree, the motility lost to damaged joints or arthritic deterioration. But while the metal and plastic implants, insofar as the materials are concerned, might 'outlive' the patient, in practice the stress and strain usually calls for renewal replacements on 7–10 year basis. And since more bone has to be removed for situating each new replacement, diminishing and finite numbers of such replacements are possible for finite human lifespans. Thus one must hope for late life, rather than mid-life cyborg parts!

Much more common, but still detachable, are many *sensory* 'prosthetics' such as optics and audio technologies (eyeglasses, contact lenses, hearing aids). Again, there has been a long history to such body-related, sensory correcting technologies. Eyeglasses were already common in Europe by the 13th century and one noted social effect was to prolong careers for scribes and accountants beyond the age when one normally needs reading glasses, thus closing off what had been in pre-eyeglass eras, jobs for younger scribes and accountants. One could continue to read into old age. Contemporary contact lenses, while still detachable, are much closer to quasi-permenent cyborg capacity. And while I have no experience of contacts, my family members do and it is clear to me that

³ *Ibid.*, p. 172.

the occasional dust or eyelash occurance, torn lenses and the like retain the sense of compromise I suggest belongs to these technologies in use. In my own case, I did not need reading glasses until age 58 when the *NY Times* and telephone directories became unreadable. These I still must use for fine print, but my distance sight remains such that no other optics are needed.

In the case of loss of hearing, hearing horns have been depicted in treatises on the senses for several centuries. But hearing horns are simply amplification devices and in most loss of hearing, more is needed than mere amplification. With normal aging, most people began to lose certain—usually high—frequencies. And these cannot be restored with amplification. In my late 60's I, too, began to notice some hearing problems and noted difficulty hearing questions from the back of the lecture halls and found cocktail party conversation hard to manage. Taking a frequency range test during a conference in Boston at the Science Museum, I found that my hearing was considerably short of the 20,000 cps younger, better hearing could detect. Today, I wear state-of-the-art, digital hearing aids and I could echo Sobchack's desire for transparency of hearing which remains much more difficult to attain with hearing technologies than in seeing with glasses. I have elsewhere described in detail the comparative embodiment processes [in the second edition of *Listening and Voice: Phenomenologies of Sound* (2007).] One of the points made in such a phenomenology is the far greater difficulty learning to re-hear, and the far greater complexity in the technologies between optics and audio devices. For example, whenever an optical prescription is changed, the user experiences very subtle changes in motility and spatiality, when walking for example, but it is but a matter of days at most before the quasi-transparency is 'fully embodied' and no longer occupies any significant role in daily life. Contrarily, the auditory embodiment in using hearing devices is much slower and more difficult to attain, so much so that many individuals give up or reject using hearing aids entirely—a fact well known to audiologists. And, even with the best and most expensive devices, feed-back at certain frequencies, the impossibility to match normal hearing's capacity to sensorily inhibit background sounds (as in a cocktail party situation), and full auditory transparency remains noticeable long after one becomes accommodated to the devices. Even one's musical memory reminds one that music no longer 'sounds the same.' Frequencies lost, remain lost, but hi-tech digital devices can partially compensate for the loss of consonant sounds—

which are often lost to those with hearing disabilities—compared to vowel sounds—which are more easily amplified, thus making speech more available than would be the case without the 'prosthesis'. They remain worth while 'trade-offs' but also they remain short of full transparency in user experience.

Closer to a cyborg body notion, implantable devices display somewhat different characteristics, and it is at this point that I shall begin with some self-reference related to "my case" and my reluctance for cyborg status:

- It began quite long time ago with dental technologies—tooth crowns—of which I now have four, plus a root canal (living tooth nerve replaced with a filler). The first broken tooth was due to a piece of sand in a Maine mussel. The silicone of the sand proved harder than my tooth enamel and so while still a graduate student, I got my first taste (!) of cyborghood. Interestingly, even though now I can still detect the differences experienced with crowns. They 'feel' different to the tongue; they lack the striated texture of an original tooth; they can cause chewing gum to stick; and the shape is never quite the same as the original tooth. While feeling-through-the-tooth as in eating has become indiscernible from the original teeth, feeling the crowned tooth with my tongue retains a discernibly different 'feel.' Thus, though permanent, the crown retains a minimal marginal self-difference. But, given either a missing tooth, or an aesthetically oddly shaped one, I would willingly choose the cyborg crown.

- Making a vast leap into animal or other human *transplants* (heart, heart valves, lungs, kidneys, etc.) I feel fortunate enough to have so far avoided this degree of cyborghood. Yet, one interesting study just now getting underway at the University of Toronto, in the HCTP group, has noted anecdotal evidence from heart transplant patients which seems to indicate that such a major transplant trauma often leads to an *experienced personality change*. Patients claim to feel as if they were 'another person' or are incorporating another person after the procedure. A carefully designed study is planned to investigate this phenomenon.

- An electronic transplant is yet another variant—pacemakers and defibrillators, for example. Pacemakers are devices, either implanted under the skin, or worn on belts with only wires implanted, which electronically 'pace' heart beats for patients whose beats are irregular or too slow. Today, some are even equiped with radio signals to a central medical facility should trouble be signaled. As with many contemporary procedures, the implantation can be performed and the patient released the following day. Most felt responses are not so much to the experience of a more regular or faster heart-beat—although this is experienced—as much as hoped for more indirect results, such as greater energy levels and less fatigue with daily activity. Some minimal training for what to be aware of is taken before hospital release. Patients describe less light-headedness, incremental energy improvement and less 'stuffiness' in their chests. Most agree that the implantation was positive enough to confirm a right choice for the procedure.

- This brings me to my own next stage cyborg addition—a medicated *stent* in one of my heart arteries. I had long delayed having a recommended colonoscopy, but finally agreed to undergo the procedure. My physician, upon doing a pre-procedure check, detected—with a stethoscope—a slight heart murmur. (It is worth noting that the stethoscope, although now an ancient device mostly used as a symbol of practicing medicine, as per television commercials, in skilled practice can reveal very nuanced internal phenomena. Auscultation, or listening-to-interiors, was a favored diagnostic art at the end of the 19th and on into the 20th century. But, as articles concerning medical education have pointed out, it is for the most part, a lost skill amongst many contemporary physicians.) So, then following the dominant *visualist* and *instrumental* practices of contemporary medicine, a series of tests was ordered: EKG, or the electrical graphing of heart motions; echocardiogram, a multi-media visualization imaging which produces dynamic images of heart motion with added graphing of the beats, *and an audio counterpart as well*. I have been researching imaging technologies now for more than a decade, so when I inform my physicians of my interest, I almost always get a detailed and interested response with demonstrations of what and how the imaging is interpreted. I also collect copies of the actual imaging performed! More, then followed, with stress tests and before-and-after imaging, and finally a recommendation to undergo an *angiogram*. This procedure is one in which a small incision is made in one's femoral artery (in one's leg alongside the

3. Aging: I don't want to be a Cyborg

groin), a catheter with fibre optic lights and internal instruments for further interventions, which is then maneuvered Nintendo-surgery style (guided by imaging on a screen and manipulated by a set of 'joy stick-like' controls) up into the arteries of the heart itself. A radioactive dye is used to show any obstructions, which when found may be forced open by 'balloon angioplasty,' that is, the obstruction is shoved aside by an inflated mini-balloon, or, if necessary, the insertion of a stent instead or in addition. The stent is a very small, wire mesh tube, often now impregnated with a slow-acting drug to prevent clotting. And, in my case, one balloon procedure was performed and one stent inserted. As with pacemaker implantation, I was released the next day and while rest was prescribed for a limited period of time, I was soon back at work.

- A phenomenology of this event is relatively simple. I was kept awake so as to move as directed; could myself see the screen which was also moved repeatedly to the advantage of the operating surgeon; feel the warming infusion of the dye, but very little pain or sensation of the catheter in motion. It seemed a quite minor interruption of life and the result was, indeed, an incremental improvement in energy levels following. I, again, have the entire procedure recorded on a DVD which I now use as a demonstration of a medical imaging technology on the road. Admittedly, it was somewhat disturbing to find—only a few weeks after my operation—that drug embedded stents were found *not to be as dramatically effective as first thought*. Some patients were found to develop clots long—several years—after the operation and thus while stenting remains more common than arterial by-pass surgery, its effects are more parallel to by-pass than previously thought. Nevertheless, had I to do it over, I would still chose the stent rather than the much more intrusive by-pass, open heart surgery! Interestingly, I have no direct bodily awareness of the stent at all—unlike my tooth crowns, it remains totally 'invisible' although the indirect awareness of energy levels may be noted. I am thus a partial *cyborg*.

By now it should be obvious that the gradual accumulation of human-technology hybridization, or the cyborg process, often relates to effects of contemporary *aging*. One's eyes gradually lose the flexibility of younger pupils, lenses and orbs; the cilia in one's

3. Aging: I don't want to be a Cyborg   39

inner ears deteriorate with age (in my case the 'boiler-maker' sounds of a farm John Deere tractor may have been my equivalent of the 'rock band' sounds of my children's generation); and plaque in one's arteries tends to build up with age. Thus, as illustrated above, cyborg strategies are often technological attempts to thwart even more severe effects of aging. I have not dealt with animal transplant strategies as fully here, but such transplantation strategies also have their associated problems, such as transplant rejection problems. Furthermore, the cyborg strategies show that such delaying tactics remain trade-offs, compromises. It is better to have a pacemaker than to have life threatening arythmia; it is better to be able to walk with either a steel-teflon implant or a prosthesis than not to walk at all; it is better to have digital hearing aids which allow seminar participation and exchange than not to be able to hear speech sufficiently to understand. Yet all these trade-off compromises fall far short of the bionic technofantasies so often projected in popular culture.

What, then, motivates the continuance of the technofantasies, the unrealistic imaginations of utopic cyborg solutions to our existential woes? At the popular level, the sheer entertainment value, mixed with the contradictory desires we have concerning our technologies may provide some of the answer. 'Explosion movies' remain popular—the quasi-super or even super powers of a "Terminator," a "Robo-cop" indulge wish fulfillments and even revenge fantasies. But they also reverberate with our secret desires to be bionic technology+human, equaling superpowerfullness. Such fantasies are, of course, ancient, but in other cultures and times did not always take technological form. To be god-like; to have animal powers; to be more-than-finite human are desires reflected in literatures and magical practices on a wide scale. Ours, reflecting the highly saturated technological texture of contemporary life, more often takes technological or cyborg form. Not satisfied with quasi-transparency; we seek the organic cyborg solution which does not happen in actuality or mundane life.

There may also exist as a motivation, a social or even industrial desire to keep utopic hopes stimulated as possible sources for technological development itself. Could, for example, technoscience development remain well funded were all utopic fantasies to disappear? I suspect that it is not accidental that Steven Hawking now urges humankind to think about escape to other planets. He is convinced that we will end up destroying ourselves and thus need to plan for extra-solar life. Yet, as physicist, he must somehow be aware that so far the only possible life supporting planet, recently discovered, lies some 20 light years away.[4] No technology existent can even approach the speed of light and thus such a journey seems close to yet another technofantasy. I am suggesting that there may be a subterranean link between the wildest science-fiction fantasies and our current technological culture. Do the desires, dreams and fantasies indirectly support our very financing of research and development?

Or, is it that the deepest desires and fantasies are simply our wishes to avoid our mortality and contingency? And, if so, is my own reluctance to cyborghood part of this same phenomenon? Should I not, contrarily, argue that precisely to accept finitude and contingency applies equally to our cyborg technologies? That is, to accept the cyborg destiny is also to accept the trade-off compromise that all our actual technologies display and this, in turn, is existentially tied to the human process of aging.

3.1 Epilogue: More cyborg than ever—open heart surgery

The angiogram with stent took place in 2006, just before I completed this article on aging for its first appearance. It is now 2008 and I have become more cyborg than ever. Once one enters today's high tech medical system, the monitoring, even surveillance process continues. By the end of 2007 I found myself often feeling highly fatigued and short of breath. So my cardiologist ordered stress tests and more echocardiograms, and later another angiogram. All revealed more obstruction to my heart arteries—again, odds, approximately half the arteries cleared by angioplasty can again close within six months to a year–and a severe 'regurgitation' with my mitral valve. My cardiologist recommended that I consider surgery and recommended a mitral valve specialist. So, looking myself at my imaging CD; turning to research on the internet; and above all investigating the specialist who was recommended by my doctor as the "rockstar" of mitral valve surgery, I ended up scheduling a consultation at Mt. Sinai Hospital's famed cardiovascular clinic. From his many entries on the 'net, which included video clips from actual operations, reprints of his articles, but above all his fervent philosophy which argued that *repair—*

---

[4] *Science*, vol. 316, 27 April 2007, p. 528.

3. Aging: I don't want to be a Cyborg   41

retaining one's own flesh, rather than *replacement* with pig or cow parts, or even a metal valve—was far superior and, where possible, yielded better results. The only machinic artifact to be involved was to install a plastic ring which would 're-seat' my valve. Here my desire to be less, rather than more cyborg kicked in.

On arrival at David Adam's office, I felt almost as I did when I played with all the cyber toys in Umea University's *Humlab* a couple of years before. There were plenty of cybertoys here as well. Adams, several of his team, awaited in a large office with a contemporary array of visual display screens. Upon these, of which there were many in the multiscreen equivalent of a newsroom, he played different takes on my latest echocardiogram and TEE, all in dynamic, colored imaging. With his cursor, he showed me that two of the chords attaching the valve to the heart were broken; he showed the backward flow of the regurgitation; and many other clearly imaged features. He confidently noted that this made me a qualified patient, and that *for him* this would be a *routine* surgery (not for me since it would be my first surgery ever). We concluded with a good discussion of the progress in imaging which characterized such high-tech surgery in the 21st century. As for risk, mortality was significantly below 1%, but with full success, quality of life could be considerable better than before or without surgery.

A bit over a month later, I entered the hospital, somewhat shocked to see that the protocol called for a *triple bypass* in addition to the mitral valve repair! I was to receive a major reconstruction. Here any phenomenology became disrupted—I had no experience of the time going into surgery until waking up in the intensive care unit, now hooked up to a large array of machines and monitors, including an external pacemaker, electrodes from my chest, and what to me appeared as a quite large leg drain from the area in which my vein had been "harvested" for the bypasses and another from my chest. I shall now cut short more details, but by the end of a week I was ready to be discharged and returned to Long Island. I was optimistic, I had had none of the frequent complications such as possible stoke, dizzy and disorienting experiences, lung congestion, or severe pain. I had only a bit of atrial defibrillation said to be common right after such surgery. And I was so relieved to have my electrodes removed, the pacemaker taken away, and to head home to the first decent food in a week. A follow-up with my cardiologist on Long Island and another echocardiogram, showed that the repair had been successful and my heart functions were excellent.

However, just in case, my cardiologist recommended I wear an "event monitor" to record any unusual rhythms from my heart, and unfortunately for me, the arrhythmia had not disappeared and a search of earlier EKGS indicated that there had been signs of pre-existing irregularities. Note here again the role of now consistent monitoring and surveillance. I had escaped major cyborgization with repair and no installed defibrillator or pacemaker, only now to wonder if I was fated to have one of these installed—I definitely did not want to be the first philosopher Dick Cheney! I felt lucky, then, when first my cardiologist, Michael Matilsky, and later a second opinion from Jeffrey Matos confirmed the earlier opinions of experts in Mt. Sinai hospital, all concurring that a pacemaker seemed unnecessary, and recommended, instead, that a heart beat regularizing medication was to be preferred. There was a catch—one must initiate the prescription only while in a 48 hour, full monitoring situation *in a hospital.* There was a slight risk of a side-effect for less than 1% of patients, that a worsening of the symptom or even a heart attack could occur. Back to both monitoring and risk taking. Fortunately, for me, nothing happened and my heart is currently very regular.

And now, the good news. Although it is only three months since my surgery, my old energies have returned; I am back to my strenuous travel schedule; and in fact feel better than I have for several years—this had been the good prognosis from such surgeries, and which could include a prolonged lifespan in an era in which lifespans are considerably longer than a century ago. My aunt who fifteen years previously, had a quadruple by-pass and celebrated her 90th birthday this last summer. Am I more cyborgian than before? Yes, and again related to the aging process, but also minimally more cyborg than could have been the case.

Here, again, one discerns the role of high tech processes, ranging from the improved imaging devices used for the monitoring and diagnosis, to the chemical technologies of medications, to the complex medical care system, but also including all the risks, assessments thereon, and trade-offs entailed with all technologies. In this case, I do not feel so bad about this particular increase in my cyborg identity.

4
Of Which Human are we Post?

Human? Posthuman? Transhuman? Did all this bother arise with Foucault? In *The Order of Things* he claims:

> Man is neither the oldest nor the most constant problem that has been posed for human knowledge... taking European culture since the 16th century...man is a recent invention within it... in the midst of all the episodes of ... that ... history [and] now perhaps drawing to a close, has made it possible for the figure of man to appear....As the archeology of our thought easily shows, man is an invention of recent date. And one perhaps nearing an end. [1]

That is, *if one accepts* a Foucault-like disjunctive-frame *episteme* account of history, *then* man – how *outré*, since feminism it must now be human – can be invented, and if invented, disinvented or deconstructed. I open in this way because the issues of the human, the posthuman and the transhuman revolve around distinctive narratives and these are often highly slippery. And, as a philosopher, I must forwarn you that I am *highly skeptical of slippery slope arguments of any kind.* At the same time, I, myself, am not unfriendly to the notion of 'posts' since I have described and others have described my own style of analysis, as *postphenomenological.*

What is the human? Biologically, modern humans, *homo sapiens sapiens,* are reckoned to be between 100,000 and 200,000 years old. How modern can you get? This is to say that biologically we differ very little from our ancient African ancestors. But is this *nature?* Not entirely. Physical anthropologists argue and recognize that many of what once would have been called cultural practices, are involved with our own human evolution. Tool use, technologies, through our pre-sapiens relatives, preceded us by more than a million years and the older arguments about how tool use, in-

---

[1] Michel Foucault, *The Order of Things* (Vintage Books, 1973), pp. 386–7.

volving complex eye-hand bodily actions are part of the way in which our brains were formed. More recently, a very provocative thesis has been put forth that the practice of *cooking* may be highly important in the evolution of our physiognomy! Cooking is a sort of 'external digestion' technology—as Ernst Kapp, the very first philosopher of technology already claimed in 1877. Such pre-digestion provides for two conditions for biological selectivities which help define modern humans: smaller teeth, a key physiological difference between us and most earlier humans, and the loss of the skull crest to which much stronger jaw muscles were attached for chewing. And, as these anthropologists also claim, cooking hearths go back to precisely to such early modern sites, but the evolutionary process begins earlier in that charcoal heaps without hearths do go back to pre-sapiens sites as indicators. I suggest that what is neat about this analysis is that it is much closer to a 'natureculture' or 'culturenature' notion as described by Donna Haraway, rather than the too clean division between nature and culture which presumably defines the 'modern settlement' a-la Bruno Latour. And it also gives a new meaning to "We are what we eat."

Or, is the modern human the human the one who was invented at the beginning of the early modern scientific era, the 17th century? This would be the Cartesian-Lockean human–the subject in the *camera obscura* mechanical body box, but individualized and a subject epistemologically, but also one who has inalienable rights to private pleasures, freedom and happiness in the social-political arena. Surely, this version of 'human' is enigmatically being called into question in a postmodern era—on the one side the notion of extreme autonomy, without social relations and net-workings, but on the other the possible loss of or weakening of civil liberties—poses an ambiguous threat to hard won enlightenment values. Can we have a less self-enclosed, less autonomous, even closer-to-the-animals human, without losing the important political gains made in modernity? The transcending of a now four century old interpretation of the 'human' is certainly timely and important.

If we are, then, at a crucial juncture, a time-warp in which we as self-interpreting animals must re-assess ourselves, then there is a type of parallelism which stretches back to the beginnings of our 'modern' era. As it turns out, this summer one of my commitments was to do a number of entries for a forthcoming Blackwell Companion to the Philosophy of Technology volume, one of which was an introduction to a section on science and technology. In the process, I returned to one of the pioneers of that modernity, Francis Bacon , who in his *Novum Organum* was aware of a turning point in his historical time and who developed the notion of four idols to be avoided in entering the new era. It occurred to me that this device could serve a good purpose for precisely this theme as well. So I shall talk us through four new idols in discussing the human, the posthuman and transhuman issues here. My idols are:

- The idol of Paradise. This is the idol of much *technofantasy* which often underlies much of the discussion context we are engaged in.

- The idol of Intelligent Design. This is the idol of a kind of arrogance connected to an overestimation of our own design abilities, also embedded in these discussions.

- The idol of the Cyborg. Cyborgs, made popular since midcentury, are hybrid creatures of human, machine, and animal combinations, but what do they imply?

- The idol of Prediction. Projections of futures are always involved in era shifts, but if past projections are taken into account, this turns out to be a very dicey practice.

4.1 The idol of Paradise

Anyone familiar with much history of the literature on paradise knows that one problem with paradise is that it is likely to be *boring*. From singing angels on a cloud to the discovery that seventy virgins may turn out to be seventy raisins due to a mistranslation, to Dante's dull paradise in the *Divine Comedy* compared to his imaginative levels of Hell, all point to the difficulty of making any utopia exciting and stimulating. I have always argued that to 'imagine up' is much harder than to 'imagine down.' Take, for example, science-fiction presentations, particularly those such as "Star Trek," "Battlestar Gallactica," and other series. When the humans and their allies, sometimes quasi-humans from other planets, come into the presence of a 'superior' set of beings, what do they find? The two most popular variants are, either these beings have superior technologies, all of which by the way can be found in ancient literatures in non-technology forms, such as powers of invisibility—now a type of electronic shield, then a cape of

4. Of Which Human are we Post?

invisibility—powers to change forms—now into a transformer or a tech exo-skeleton, then into a dragon or a spider—and so on. Or, the superior ones have extraordinary spiritual or mind capacities. The can see futures, meld with other minds, communicate without technologies, and are usually peaceful meditator types. From flying carpets to warp speed, I note there is little new in such fantasies. The difference is that since modernity the fantasy embodiments have tended to be *technological* rather than organic, animal-like, or supernatural. Contrast Bruegel's organic animal, supernatural tormentors in his paintings with da Vinci's fantasy technologies and you get something of the shift.

I have earlier argued that fantasies take shape and form in relation to the relative lifeworlds of the inhabitants.[2] Thus, if one lives in a world in which daily life includes frequent and existentially important interactions with animals, and for that matter, plants, as in hunter-gatherer cultures, then the wish fulfillment fantasies will take the shapes of animal fantasies, dreams, stories, or of plant cycles and growth and decay metaphors. But, if your lifeworld is one saturated with a technological texture, then you get the more 'modern' versions suggested above. The technologies will provide the magic answers. Our myths are indexed to our experiences.

Clearly the implication is that our current debates concerning human/posthuman/transhuman take this current techno-mythological shape. Let me begin minimalistically with enhancement desires: Do we want more muscle power? Bigger breasts, fuller lips or tighter buttocks? Larger penises or better erections? Steroids, breast implants, Botox, liposuction or tucks, penis surgery or Viagra—this drill,is apparent on television, 'net spam [ironically, my wife gets more penis enlargement spam than I do]. This has led to changes in time awaiting the doctor in which time is shorter for Botox injections than wart removal, in turn this is related to capitalism in the sense that injections are profitable and easy and wart removal is limited by insurance, all this is part of "Modern Times" in the post-Charlie Chaplin movie we live in. All these techniques *work*–but not without unintended consequences. Steroids increase the risk of early heart problems; silicone implants can leak and seem to be implicated with auto-immune diseases; long term Botox use has toxic effects, and in the wrong mix Viagra can cause critical low blood pressure or blindness. Paradise is not to be found

---

[2] See my "Technology and Human Self-Interpretation," *Existential Technics* (SUNY Press, 1983).

4. Of Which Human are we Post? 47

here; calculated risk and trade-off compromises are to be found here.

Here, then, is my thesis: the desires and fantasies are ancient. Historically, they appear in our literatures, our fairy tales, our art. The fantasies and desires then want some *kind of magic* to fulfill the desire-fantasy. But the form of the magic differs, according to my thesis, by the textural patterns of historic lifeworlds. So, from magic potions to magic injections, from an age of alchemy to one of chemistry, the fulfillment technique will differ. But, why do I call it magic? Because magic, unlike actual chemistries and technologies, does not have ambiguous or unintended or contingent consequences—trade-offs are lacking, only the paradisical results are desired.

How, then, does this relate to the human-posthuman-transhuman discussion? The answer is simple in one respect—to locate the desire-fantasy, look for the *hype*. Technofantasy hype is the current code for magic. Switch examples now from personal enhancement desires to technologies which will fulfill our social energy desires—remember the time when the world community, fearful of a nuclear holocaust, hyped the 'magical' transformation of that power into peaceful uses? One was the technofantasy of limitless, almost free nuclear supplied power.

This is precisely an example of magical thinking, the hype which projects a non-contingent, non-consequential, non-trade-off solution. The 'infinitely inexpensive' projection did not take into account the need for safety reduncancy, for security factors, and for the still problematic need for hundreds of thousands of years safe waste storage, all of which complicate 'paradise,' and all of which need to be calculated into the *costs* of this non-neutral technology. Please note here that I am not arguing a dystopian view—it may be possible with very careful planning, with contingency considerations and new technologies to make such an energy source less long range dangerous than it now is. Rather, I am arguing that magical thinking disregards the ambiguous, non-neutral character of actual technologies. Desire-fantasy, with respect to technologies harbor an internal contradiction. On the one side, we want the super-powers or enhancements which technologies can confer—long range vision with telescopes, mountain moving capacity with earth movers, supersonic speed with jet power—but on the other, the technofantasy is to have this enhancement be so totally transparent that it *becomes us*. This is a Superman technofantasy; to have and to *be* the power embodied. Such then are the dynamics behind the idol of Paradise.

4.2 The idol of Intelligent Design

Most of you are familiar with this term from the currently raging evolution/creationist debates popular in the United States. In that context, "intelligent design" is the notion that various natural phenomena, particularly forms of life, are too complex *not* to be intelligently designed. The implication, of course, is a throw-back to the old teleological argument for God, that smart design implies a smart designer. Now, were I to plunge into the evolution/creationsist argument, I would in my usually perverse provocative way, probably *invert* the proposition and argue that evolutionary results are *in fact too complicated to have been designed!* And I would look at the current state of robotics as a very good illustration of this inversion. To date there are no robots with the gracile motility of even insects-in-motion, let alone simulacra of upright posture humans playing tennis. Beetles are better negotiating chaotic terrain than robots and in terms of flight, bumblebees and humming birds make mockery of the smart bombing "Predator" of Iraq War fame. Once again, let me warn you that my ironic gestures against this sense of intelligent design are not indicators of a lack of appreciation for technological innovation and modeling-simulation experiments. To the contrary, one of the most delightful and amusing of my observational side-lines over the years has been to witness the way a lone phenomenologist, Hubert Dreyfus, so provocatively influenced the trajectory of both AI (artificial intelligence) and robotics.

His application of Heidegger, picked up by Terry Winograd and Samuel Flores, in programs called 'ontological design' changed office systems to much more user-friendly platforms, but in robotics, the Merleau-Pontean notion that bodily motility underlies all intelligent behavior has deflected design notions regarding robot motility. For example, the old dicta regarding a central nervous system centralized in a sort of brain-in-a-vat model for deciding and directing robot motion, has gradually begun to be replaced by 'smart insect' models of less self-conscious motility leading to better abilities to locate obstacles and such. Both directions are bodily-being-in-environment models with greater reliance upon perceptual analogs than upon calculation machine capacities. Perhaps embodied beings are less calculational machines and more sentient animals than modernity usually thinks?

Permit now a shift of example. In this case human intelligent designers, recognizing the gracile motility of our fellow beings and in line with the previous desire-fantasy dreams, now wish to fulfill the ancient desire for *flight*. As I have suggested, the earliest stages of modernity began to shape such desires into technological forms, and whereas most of the imaginations pre-renaissance used large birds, dragons or other flying animals—or sometimes out-of-body dream flights—Leonardo da Vinci began to visualize different flight technologies. Some were quite naively amusing, such as his presumed anticipation of a helicopter, a 'flying screw' machine which, of course, could not possibly work! But, he was also an avid observer of birds and birds have always been icons for the human desire to fly. And Leonardo was a keen enough observer to note that bird wings contain curvatures in form which we now know allows for *lift*—and he incorporated this into his drawings of winged flight machines. None, again, could have worked. But why? Some have argued that the conceptual design was good, but the lack of light weight materials and the lack of tensile strength of materials prevented such possibilities. Indeed, when I made remarks of this sort not long ago in a review in *Nature*, I was taken to task by an editor who pointed out that a designer inspired by da Vinci, had indeed built a hang-glider along deVinci lines, which did glide. However, when I examined this design, I discovered that a whole series of design modifications totally unknown to da Vinci had been incorporated. But in either case, once again the properties and capacities of the technologies needed to be taken into account.

Humans can be stubborn, so the dream of human powered flight persisted. We look back at the funny home movies of the clumsy attempts at flight technologies at the end of the 19th century, and when flight succeeds—with modified bicycle parts and finally a non-human engine, in 1903 and the Wright flight—a quite different trajectory is born. The finally successful powered flight was actually a combination of many hybrid technologies, light and flexible, strong materials, control designs for fixed wings, and a small internal combustion engine and propeller (a variation on the ancient screw machine), and an abandonment of the bionic 'bird model' of da Vinci. Rather, the developmental history points to what Andrew Pickering calls the "dance of agency." That is through much human-material interaction, from which emerge new design and trajectory factors, led to today's very non-animal like flight. So, when finally one successful human-powered aircraft does appear, the "Gossamer Albatross", powered by a highly trained bicycle racer with similar technology driving a large, slow propeller, in a mylar-plastic airframe, which flew across the English Channel in 1979, the stubborn fantasy was fulfilled. Yet fulfilled only in ideal, limited conditions and with the appearance of a sort of clumsy, anachronistic success.

Once again the idol of intelligent design gives way to a human-material or human-technology set of interactions which through experience and over time yield to emergent trajectories with often unexpected results. The fantasy model of an intelligent, autonomous designer, working out an intended result upon a purely 'plastic' material, yielded the proper result gives way to the more realistic notion of human-material interaction, through experienced 'resistances and accommodations' in a 'dance of agency' a-la Pickering, or the invention of an entirely new set of uses for a useless 'glue' as in Latour's description of the Post-it.[3]

4.3 The idol of the Cyborg

Although it was probably Donna Haraway who made the figure of the cyborg into its best known form, as the non-innocent hybrid of human-animal-and machine moving amidst the techno-science naturecultures of postmodernism, the cyborg was gestated in the *cold-think* of World War II and then the Cold War. From Clynes to Wiener to von Neumann to Herman Kahn, the technofantasies of moving beyond the humanistic were configured. First, in the Manhattan Project and thinking the unthinkable, then on to its 'peaceful' uses—such as creating huge atom-bomb produced harbors in Alaska—cold-think prided itself on machinic thinking replacing human thinking. One of the main technologies of cybernetics, after all, was to create a non-evadable aircraft artillery fire.

But the slippery slope fantasies are perhaps better seen when science-fiction and its filmic expressions are introduced, as in "Terminator," "Robo-Cop," and the variations upon 'bionic' men and women. It is here that a history and phenomenology of *prostheses* can be informative: Prosthetic replacements for limbs and other body parts have an ancient history. Wooden teeth and detach-

---

[3] Bruno Latour, *Science in Action* (Harvard University Press, 1987), p. 140.

able artificial limbs go back to ancient mummies. In experienced use, these prostheses fall into what I have earlier called *embodiment relations,* that is, we humans can use technologies through which we can experience our environment by 'embodying' such devices, and while in use such devices are 'experienced through' in a partial transparency or partial withdrawal. We do not attend to our eyeglasses, or better, our contacts; Merleau-Ponty's lady with the feathered hat or the blind man with a cane, can 'feel' through these extensions for bodily motility in an environment. *But,* the withdrawal or transparency comes with both a partial incompleteness, and more, with a selectivity such that what is experienced through the prosthesis is both magnified in some aspects and reduced in others. The peg leg, or its high tech, Iraqi War hydraulic replacement leg cannot 'feel' the hot, sun-baked surface of the sidewalk the way one's bare foot can. But through the prosthesis one might be even more sensitive to slipperiness or rough texture. Once again, it is the sensitivity to the materiality of the prosthesis which slippery slope fantasies forget. Prostheses are compromises; we may have them, but we fall short of experiencing a total transparent embodiment. At a very low and simple level, with a tooth crown, there may be a very high transparency, we are rarely aware of which tooth is crowned. But at a more complex level—say hearing aids—it becomes obvious that transparency is at best partial. In my own case, such painful occasions ranging from dinner parties to a bar, the background noise cannot be dampened even with my hi-tech digitals and remote with ambient sound suppressing programs. Nor is music, either live, or worse on the radio, what I can remember it once being.

Returning to limb prostheses, today the attempt to have the artificial limb mimic likeness to the original or missing limb, has sometimes given way to a different variation entirely. In a trajectory away from similarity and away from the contradictory having and not-having a technological self is the move to have a different kind of prosthesis. Aimee Mullins, who played 'Cheetah Woman' in *"Cremaster 3",* learned to use from childhood a set of spring like legs. She was born with fibular hemimelia, a birth defect born without fibula bones, and underwent stump amputation at age one. The spring legs, subsequently used by a number of athletes with the same defect, are *not like* human limbs, but give a selectivity which magnifies spring-powered speed capacities. Oscar Pistorius, a South African with spring powered transtibial prosthetic legs runs almost as fast as normally legged runners, but was denied Olympic entry in part because some feared such hi-tech devices might give an advantage. This is a trajectory which, while not returning our sentient bodies to us, allows us different capacities than before.

Even more internalized, knee, hip, shoulder and other *implants*, like Viagra and Botox, work and work better than damaged parts, *but implants, unlike the fantasized eternality of perfect machines,* also wear out! Metal and plastic 'ages' and must be replaced every seven or more years, and because more bone must be cut away for the replacement part, leads to diminishing returns. Here, again, is contingency and trade-offs, and it is better not to have to undergo such procedures unless necessary and hopefully at older ages. The cyborg, when critically examined with a concern for its materiality, does not display its science-fiction technofantasy form. The cyborg, too, can be an idol.

4.4 The idol of Prediction

In the same narratives concerning the human, the posthuman and the transhuman, both dystopian and utopian predictions produce idolatrous technofantasies. Here I could wax eloquent for pages, but I select a few predictions some of which are made by prominent scientists, others by those extolling utopic virtues in magazines, most selected for deliberate irony by current lights:

- Lord Kelvin, 1895, "Heavier-than-air flying machines are impossible."

- Ken Olson, 1977, of Digital Equipment, "There is no reason anyone would want a computer in their home."

- Recall the above cited infinitely cheap atomic energy prediction—this era included other predictions including the atomic car which could go 500,000 miles without refueling.

- Or extolling modern magical materials such as the beautification of walls with white lead paint; the amazing material, asbestos, for floor tiles, roofs, insulation and decorative interiors.

- How about radioactive suppositories? So every tissue in the body could benefit from healthful radiation.

4. Of Which Human are we Post?   53

These are examples from *Follies of Science*, (2007)[4] these can more than be matched with lists from Tenner's *Why Things Bite Back* (1996) when he cites Toffler's famous prediction of how the electronic society would be the 'paperless society,' and how home security systems, by generating false alarms tied down the equivalent of 58 police officers full time answering 157,000 calls when only 3,000 were genuine, thus likely diverting attention to other crimes. And the list can go on.[5]

David Nye, in *Technology Matters* (2006) points to an in depth survey of predicted technologies, 1890-1940, 1500 predictions, and less than one-third occurred. This, by the way, concerned what technologies would be invented, not what uses, unintended consequences, what reversals would occur. Chiding me for pointing this out in a *Nature* and claiming these are pretty good odds, my response is that 50% odds are normal for a penny toss, these are less than that!

Now, you will note that I have not addressed many of the famous predictions coming from post- and transhumanists, for example, those issued by Hans Moravek , concerning downloading a human mind into a computer, and Ray Kurzweil, concerning the age of intelligent machines. Do these worshippers of the idol of prediction have credibility? Pause for a moment: Just what of a human mind would, should, could be downloaded? The internet, which plays a strong role in Kurzweil's fantasies, turned out to have some avery unpredictable outcomes in relation to its original design and intent. As everyone knows, the decentralization and distributive network technologies of the early internet—largely restricted to the Cold War university cold-thinkers and the military—was designed to be non-defeasible by nuclear attacks. No central authority or power could overcome the distributed networks. Yet, once access was expanded—and is still expanding—this lack of central 'nervous system' analog control has led to all sorts of unintended consequences. From Andy Feenberg's analysis of the French minitel with its dating game results, to the current American obsession with hide and seek and avatar sting operations for pedophiles, non-defeasibility has turned out to have lots of unintended consequences. Thus, Kurzweil's almost accidentally correct prediction than the Soviet Empire would fail due to the corrosive power of

---

[4] Eric Dregni and Jonathan Dregni, *The Follies of Science* (Speck Press, 2007).
[5] Edward Tenner, *Why Things Bite Back* (Alfred A. Knopf, 1996), p. 7.

rising internet-communication-distributed networks did point up the potentially democratic effect, subversively democratic effect of this technological complex. But, then, turn this back to Moravek's notion of downloading the human mind into a computer—and by extension onto the internet—and what do we have? Were we all merely cold-thinkers as per von Neumann and Herman Kahn, this might be bad enough, but how about pedophiles, and all the rest of the Freudian 'unconscious' aspects of the human mind, downloaded and distributive through the 'net? What does it mean to download a mind? If it means downloading all the 'bad' parts along with the 'good' parts, are we not back at the copying machine? Which is, after all, the perfect reading machine which faithfully reproduces precisely the page it is given? Were Moravek himself downloaded, would he be any better than he now is? And, if not, are we stuck with a possibly flawed Moravek now forever?

Note here that my worries are **not** at all those of romantics, objecting because this is 'unnatural,' nor are they those of the theistically inclined, concerns with human hubris, overreaching our natural human limits. They are, rather, worries about unintended consequences, unpredictability, and the introduction of disruptions into an ever growing and more complex system. They are worries about how 'normal accidents' get built into systems as per Charles Perrow. And, my worries arise precisely from what I have learned about technologies in the now nearly four decades of thinking about technologies. My worries focus upon precisely the *disregard* for the materiality of technologies, the ambiguity of technologies, the multistability of technologies, and above all the intimate role of *humans with technologies*. Thus I will conclude with another narrative which I hope will capture the sense of what I have been talking about.

4.5 John Henry and Big Blue

The American 'John Henry' legend expressed in songs and tales reflects an earlier era in which technologization was feared with respect to replacing humans, but in this case laboring humans. John Henry was depicted as a big Black man, known for his exceptional skills at driving spikes for setting up rails for the advancing railroad—in one version, in another he was depicted as a tunnel digger. In both cases an invention of a steam powered spike driver in the first, or a steam powered digger in the second, threatened

4. Of Which Human are we Post? 55

to outdo and replace John Henry. So, a contest is set up between John Henry and the steam machine and with superhuman effort, John Henry scores a very tight victory over the steam machine—but his efforts ended with a heart attack and he collapsed at the finish line, dead.

Of course, we know the outcome, the machine actually wins, since once automated, driving spikes or digging tunnels, steam machines replace human muscle power—and as the moral of the story goes for labor unions and a social left, the armies of Coolies who did that work on our 19th century railroads, were left unemployed. But, fast forward, who today bemoans the replacement of hard, 'chain-gang-like' labor with efficient machines? I switch to my observed version from two events at my retreat in the Green Mountains of Vermont. Now almost a decade ago, a devastating ice storm from Canada coated the forests of my region and mountainside, damaging many trees and downing others. I have a managed forest plan and on advice of my forester, accepted a selective cut. The end result, seven double truck loads of logs were cut and sold, and the stumps cut down to the ground and brush burned or removed. And how did this happen? With one tough old Vermonter, armed with a large chain and four wheel machine, complete with road making blade, chain saw in hand *by himself*—no, *by himself plus his technologies*—in a matter of weeks completed the task. Then, again this summer, this time wanting selective trees which had grown up in my lower meadow and apple orchard, now threatening my highly taxed view—there is a view tax in Vermont!—I had my meadow mowing Vermonter do the job. This time, with a large excavator with clasping arm, dozer blade and another of the large four wheel machines—again by himself, no, by himself with the large machines—does the job in a few days. I imagined a century ago when both these jobs would have called for a gang of Vermonters, horses and sledges, and hand powered two-man saws, undertaking what would have been a month's work. My point? The technologies *did not* replace the humans; rather different technologies *plus* the humans changed the nature of the task.

Today, here in the context of the human, posthuman, transhuman narratives the variant is, once again, the humans *versus* the machines, this time not with respect to muscle power, but with respect to *calculating power*. AI, VR, and the range of more 'mind' related technologies are again mythologizing the human versus machine myths long embedded in our culture. Now, frankly, were my computer able to simply ingest all my tax related data for my annual income tax report, and spit out a legal and yet maximal result for me, I would cheer and accept giving up the task entirely. That is, of course, not the way it happens. Instead, it happens—if I borrow from Latour's human-and-non-human collectives notion- more like this: I collect and organize my annual data, which now enormous and complicated as it is, and turn it over to my tax advisor. He, now with four others in his office, format it for the programs which are possible for analysis and last year he said he tried some seven variants to produce the most effective result. This is a simulation and modeling process now so common for complex phenomena problems for which such calculation machines are at their best. But, again, it is clearly not human versus machine, it is humans in conjunction of machines which produce the result.

And this is where I finally turn to my last legend, the 1997 presumed *defeat* of champion chess player, Kasperov, by the IBM computer with the Big Blue program. The PR—and the Minsky's and Moraveks and all the other technofantasizers—hyped this occasion as the ultimate, inevitable result of yet another mythical 'machine-beats-human' contest, a mental and century later version of "John Henry." But, that is not what happened and the history of the event not only is different from its mythical version, but precisely needs to be reframed in human *plus* machine interpretation. From the first, of course, it is human plus machine in the creation of the software—the software did not create itself, it was honed and refined by many skilled programmers, as per the previous tales in amongst the idols, and gradually perfected through resistances and accommodations and the dance of agency peculiar to computer programming. But there is more: during the match, after each game, but behind the scenes, somewhat like the gang of water dabbers and cleaners at a boxing match, Big Blue was aided by its programmers who tweaked and re-tweaked its programs before the next round. This was not machine versus Kasperov, this was the collective machine plus programmers, a collective versus Kasperov! Is it then any wonder that Kasperov is as much exasperated by the behavior of the 'machine' as he is by the lightening quick moves it can make with hyperspeed calculations?

I suggest here, that only if humans are stupid enough to end up worshiping the very idols they create, could the fantasized replacement of humans by machines take place. Rather, the changing technologies with which we interact, form collectives, experience the dances of agencies, do forecast vastly changed conditions of work and play and even love, but it is not them versus us. In Long Island my living room has a number of pieces of 'art' from the Sepic River region of New Guinea. I bought these pieces while in Australia, from a shop in Sydney which specialized in this sculpture-'art.' Now, in their own cultural context, such pieces are not at all what we would think of as 'art' but were simultaneously more a sort of 'practical religious' set of objects. They served fertility, ritual, healing and many other social functions, they were what older anthropologists might have called 'sacred' objects—or *idols*. But, if they are sacred, how could I acquire them? The answer is one which I find appropriate for my conclusion here. These sacred objects, idols, are in their original context, thought to gradually lose power, to deteriorate, even to break-down—*amazingly just like technologies*—so when they reach a certain stage of uselessness, they are discarded. And so I have collected some of these discarded idols and re-formulated their use into 'art objects' in my home. Here lies the moral of my tale concerning the human, the posthuman and the transhuman.

CPSIA information can be obtained
at www.ICGtesting.com
Printed in the USA
LVHW02s0320140818
586919LV00001B/8/P